Handbook of organic solvent properties

Handbook of organic solvent properties

Ian M. Smallwood
Consultant

ARNOLD
A member of the Hodder Headline Group
LONDON • SYDNEY • AUCKLAND
Copublished in the Americas by Halsted Press
an imprint of John Wiley & Sons Inc.
NEW YORK • TORONTO

First published in Great Britain 1996 by
Arnold, a member of the Hodder Headline Group,
338 Euston Road, London NW1 3BH

Copublished in the Americas by Halsted Press
an imprint of John Wiley & Sons Inc.,
605 Third Avenue, New York, NY 10158

British Library Cataloguing in Publication Data
A catalogue record for this book is available from the British Library

Library of Congress Cataloging-in-Publication Data
A catalog record for this book is available from the Library of Congress

ISBN 0 340 64578 4
ISBN 0 470 23608 6 in the Americas only

Produced by Gray Publishing, Tunbridge Wells, Kent
Printed in Great Britain by The Bath Press, Avon and
bound by Hunter & Foulis Ltd, Edinburgh

Contents

Introduction vii

Key to tables ix

Section 1 **Hydrocarbons** **1**
- *n*-Pentane 3
- *n*-Hexane 7
- *n*-Heptane 11
- *n*-Octane 15
- *n*-Nonane 19
- *n*-Decane 23
- 2,2,4-Trimethyl pentane 27
- Cyclohexane 31
- Benzene 35
- Toluene 39
- Ethylbenzene 43
- Xylene (mixed isomers) 47
- C_9 Aromatics 51
- Tetralin 55

Section 2 **Alcohols** **59**
- Methanol 61
- Ethanol 65
- *n*-Propanol 69
- *i*-Propanol 73
- *n*-Butanol 77
- *i*-Butanol 81
- *s*-Butanol 85
- *n*-Amyl alcohol 89
- *i*-Amyl alcohol 93
- Cyclohexanol 97
- *n*-Octanol 101
- Ethanediol 105
- Diethylene glycol 109
- 1,2-Propanediol 113

Section 3 **Glycol ethers** **117**
- Propylene glycol methyl ether 119
- Ethylene glycol methyl ether 123
- Ethylene glycol ethyl ether 127
- Ethylene glycol monobutyl ether 131

Section 4 **Chlorinated solvents** **135**
- Methylene chloride 137
- Chloroform 141
- Carbon tetrachloride 145
- 1,2-Dichloroethane 149
- 1,1,1-Trichloroethane 153
- Trichloroethylene 157
- Perchloroethylene 161
- Monochlorobenzene 165

Section 5 **Ketones** **169**
- Acetone 171
- Methyl ethyl ketone 175
- Methyl isobutyl ketone 179
- Cyclohexanone 183
- *n*-Methyl-2-pyrrolidone 187
- Acetophenone 191

Section 6 **Ethers** **195**
- Diethyl ether 197
- Diisopropyl ether 201
- Dibutyl ether 205
- Methyl tert butyl ether 209
- 1,4-Dioxane 213
- Tetrahydrofuran 217

Section 7 **Esters** **221**
- Methyl acetate 223
- Ethyl acetate 227
- Isopropyl acetate 231
- *n*-Butyl acetate 235
- Cellosolve acetate 239

Section 8 **Miscellaneous solvents** **243**
- Dimethylformamide 245
- Dimethylacetamide 249
- Dimethylsulphoxide 253
- Sulfolane 257
- Carbon disulphide 261
- Acetic acid 265
- Aniline 269
- Nitrobenzene 273
- Morpholine 277
- Pyridine 281
- 2-Nitropropane 285
- Acetonitrile 289
- Furfuraldehyde 293
- Phenol 297
- Water 301

Bibliography 305

Introduction

It goes without saying that a solvent will be chosen to do its job effectively and economically, and it is usually possible to choose a short list of solvents which will do this without requiring a large amount of experimental work provided that reference books, setting out the properties of commonly used solvents, are available.

Almost always the choice will lean towards a solvent that is already used on the site, or to one of which the researcher has experience.

Today, however, there are other criteria than solvent power and volatility which need to be considered. Regulations covering the exposure to solvent vapours of makers of the product and its users are much stricter than in times past and, as knowledge of the potential dangers improve, are likely to become even stricter. No longer is it possible to protect the makers and users by improved ventilation to draw the solvent fumes away and to discharge them, heavily diluted, into the atmosphere.

Solvents' effects on both high- and low-level ozone in the atmosphere are now unacceptable; although it is not widely appreciated by the general public that solvents contribute a large part of the volatile organic compounds (VOCs) in European industrial countries, as much as the whole arisings from road transport uses.

This has a significant influence on solvent choice, since any solvents that evaporate in industrial operations have either to be recaptured or destroyed rather than passing unchanged into the atmosphere. The economics of solvent choice may now allow an expensive solvent to be used many times over rather than a cheap one only once, always provided that the expensive one can be recovered in a fit state for reuse. A low-cost solvent may be difficult to destroy by incineration in an environmentally acceptable way, perhaps because its molecule contains chlorine, nitrogen or sulphur so that its disposal cost may exceed its purchase cost.

Another factor of considerable importance is the need to avoid changing the solvent to be used in a process. This even applies to the earliest stages of the development of a new product since the temptation to stay with a solvent that appears to be working well in the laboratory is great. The longer during the development stages that toxicity, environmental damage and overall economics are not considered in detail, the more difficult it is to make a change.

Once production is started long and difficult negotiations with regulatory bodies, which often need to re-approve an altered process, may be involved and a change of solvent becomes almost impossible.

All these considerations make the optimum selection of a solvent for a process a matter of importance. Fortunately much information is available in the literature concerning both the properties of the old solvents (e.g. benzene, carbon tetrachloride) which were often by-products of other processes and of the newer ones (e.g. tetrahydrofuran, dimethylacetamide) which are purpose-made for their desirable solvent effects.

This book is a collection of the physical properties of most commonly used solvents along with information on their behaviour in the environment during and after use and their health and fire hazards.

Key to tables

Name

It cannot be stressed too strongly that the name of a solvent should be easily used and recognized by all, from the graduate in the research laboratory to the plant operator who may have difficulty in reading a language which is not his or her own. Once a name, or worse still a set of initials, has become standard usage on a site it is very difficult to make a change.

The use, for instance, of IPA for isopropanol or isopropyl acetate of tri or TC for trichloroethylene or 1,1,1-trichloroethane can lead to errors that are very serious. Highly toxic benzene can all too easily be confused, in dealings with Europe, for benzin, a comparatively low toxicity material.

Hazchem code

This is a code informing U.K. emergency services of the action to be taken when dealing with transport emergencies and can be a useful method of labelling storage tanks on a site where many different solvents are handled. It consists of a number and one or two letters (see Table 1).

Explanation:

- It can be seen that breathing apparatus (BA) should be available in all cases.
- E after the code indicates that evacuation of people should be considered.
- 'Contain' means that any spillage should not enter water courses or drains.
- 'Dilute' means that a spillage should be washed away to drain with plenty of water.

Table 1 Hazchem codes

Number	*Firefighting medium*
1	Water jets
2	Water fog
3	Foam
4	Dry agent

Letter	*Explosion risk*	*Personal protection*	*Action*
P	Yes	BA + Full	Dilute
R	No	BA + Full	Dilute
S	Yes	BA + Gloves	Dilute
(S)	Yes	BA (fire only) + Gloves	Dilute
T	No	BA + Gloves	Dilute
(T)	No	BA (fire only) + Gloves	Dilute
W	Yes	BA + Full	Contain
X	No	BA + Full	Contain
Y	Yes	BA + Gloves	Contain
(Y)	Yes	BA (fire only) + Gloves	Contain
Z	No	BA + Gloves	Contain
(Z)	No	BA (fire only) + Gloves	Contain

Molecular weight

On many occasions the effectiveness of a solvent will be compared on a molar rather than on a weight or volume basis.

Purchases are, however, always by weight or by volume so that a low molecular weight solvent may have a significant cost advantage in use. On the other hand the low molecular weight of water, present in all solvents at parts per million level at least, may be surprisingly damaging when processing, for instance, high molecular weight Grignard reagents or urethanes. When a solvent is used at a high mole fraction, as for instance in extractive distillation where solvent mole fractions of 0.9 are common, the cost-effectiveness of a low molecular weight solvent such as monoethylene glycol can be remarkable in comparison with some other entrainers.

Boiling point

Many operations with solvents involve boiling the liquid solvent and this requires a heating medium (hot oil or steam) at a temperature 15 or 20°C above the solvent's boiling point. It should be borne in mind that some solvents (e.g. DMF and DMSO) are not stable at their atmospheric boiling points and if necessary must be boiled at reduced pressure.

The normal factory steam pressure is about 10 bar and this should yield a temperature of 160°C at the point of use and boil a solvent at 140–145°C. If a higher temperature than this is necessary hot oil, stable to 300/320°C, will provide heat usable at 270/280°C. A solvent in which an involatile solute is dissolved will boil at a higher temperature than the pure material. Typically, the boiling point will be raised from 140 to 150°C if the mole fraction of the solvent in the mixture is reduced by 20%.

If solvents need to be separated by distillation it is not a reliable guide to assume that because their boiling points are widely different the split will be easy, particularly when water may be present.

Freezing point

Several solvents (e.g. dimethyl sulphoxide and cyclohexanol) are solid at ambient temperature and therefore need to be stored and handled in heated storage and pipelines and particularly with heated tank vents.

For certain materials, such as benzene, that combine a high freezing point with high toxicity, the thawing of blocked pipelines can be a difficult and potentially dangerous task. It should be noted that some solvents even when solid give off an explosive vapour. Thus, the vapour pressure of solid benzene is given by

$$\log_{10} p \text{ (mmHg)} = 9.85 - 2309/T$$

and the concentration of benzene at 0°C is 7600 ppm, which is well above its lower explosive limit.

Air-cooled condensers can be severely damaged if some of their tubes become blocked while others are still handling hot vapour causing high stresses in the tube bundle. Drums of solid flammable solvents pose handling and emptying problems.

Specific gravity

Storage tanks and their surrounding bunds are normally tested using water and are designed for a liquid of density 1.0. While most solvents have specific gravities below this, the chlorinated solvents are much denser (e.g. perchloroethylene, specific gravity 1.62) and tanks may need to be derated if switched to storing such materials.

For the same reason 200-litre drums of chlorinated solvents may be too heavy to handle, either manually or palletized on a fork-lift truck and existing pumps may be overloaded.

On the other hand, a change to a less dense solvent may mean that a full tanker load of solvent cannot be accommodated in an existing tank built with a denser solvent in mind.

When compositions are quoted as percentages it is important to know whether these are by mole, weight-by-weight (w/w), volume-by-volume (v/v) or weight-by-volume (w/v) and, in the last case to appreciate that the sum of the components will not add up to 100.

Liquid expansion coefficient

Organic solvents have an expansion coefficient five to seven times greater than water. The increase in volume when a high boiling solvent is heated from cold to its boiling point is significant and has been known to cause damage in batch-still operations when sufficient ullage has not been allowed.

When purchasing solvents by volume rather than weight it may be necessary to use temperature correction.

Flash point and explosive limits

The lower explosive limit (LEL) of a solvent corresponds to the vapour concentration above the liquid at its flash point at which a source of ignition will set off an explosion. The upper explosive limit (UEL) is the vapour concentration that is just too rich to explode and an 'upper flash point' of a pure solvent can be calculated if the UEL and the Antoine constants are known.

If the atmosphere should be enriched with oxygen it will form an explosive mixture over a wider range than that between LEL and UEL and if the 'air' is less than 8–10% oxygen, depending on the solvent, no explosion can take place at any solvent content. It is common when the likely ambient temperature lies in the range between the two explosive limits to blanket the vapour space in a storage tank with inert gas. For safety a gas with about 3% oxygen is used, but it should be remembered that this 'nitrogen' is not free of oxygen so that solvents that form peroxides very readily (e.g. ethers) can be damaged if pure nitrogen is not used as the blanket over them.

Testing for flash point is carried out using laboratory equipment of a range of designs. Those that are easier to use do not necessarily correspond to the standard test methods laid down by regulatory authorities, but are adequate for internal purposes on site. The standard methods are difficult to use for an inexperienced operator, but all common mistakes tend to give a test result lower than it should be and therefore err on the side of safety. The figures quoted here are ones using the tag closed cup method which tend to give rather lower results than the tag open cup and the Cleveland open cup methods. There is no reliable conversion factor between the various methods.

Mixtures of two or more solvents may, because of the presence of azeotropes, have a lower flash point than their components have separately.

When handling solvent-laden air, as is common in activated charcoal recovery plants, it is normal to operate with a flammable solvent content in the range of 25–40% of LEL. If information on the flash point of a mixture is not available the great majority of solvents have an LEL of 10,000 ppm (1%) with a few in the range of 7000–10,000 ppm. The flash point of straight run hydrocarbon solvents (e.g. white spirit) can be estimated from their initial boiling point (IBP)

$$\text{flash point} = 0.73 \times \text{IBP} - 72.6$$

where both temperatures are in °C. Common practice in United States is to quote petroleum temperatures in °F.

Autoignition temperature

While generally a spark or flame is needed to set a flammable liquid on fire, almost all solvents can be ignited by a very hot surface and some by heat sources that are commonly met on industrial sites such as steam mains, hot oil pipelines and items heated by electricity, including laboratory heating mantles.

Steam pipes routinely have temperatures between 160 and 200°C and may be considerably hotter where high pressure steam is used. Hot oil reaches 300°C or a little higher. Solvents such as ether with an autoignition temperature of 160°C and dioxane (180°C) are therefore liable to catch fire if dripped on to a heating medium line. Their use on a site may require major changes to plant layout. Carbon disulphide has an autoignition temperature of 100°C and cannot safely be used except in a purpose-built plant. The glycol ethers also present a hazard when hot oil heating is used.

Electrical apparatus that is correctly described as flameproof can reach the autoignition points of some solvents and a change of solvent in a manufacturing facility should not take place without this being considered.

Electrical conductivity

When solvents are moved in contact with another phase static electricity is generated. This can occur in a number of circumstances in industrial operations such as

1. A hydrocarbon/water mixture is pumped in a pipe.
2. A solvent is stirred or pumped in contact with a powder.
3. A solvent is sprayed into air.
4. A solvent is contacted with an immiscible liquid (e.g. water) in an agitator.

If the static produces a spark which contains enough energy and if the vapour phase in contact with the liquid is between its LEL and UEL an explosion may occur. It is also possible that a fine mist of flammable liquid below its LEL can be ignited by a static spark.

The chance of such an explosion depends largely on the electrical conductivity of the solvent (see also the section on *Dipole moment*) since a high conductivity allows the charge to leak away. Some solvents have naturally high conductivities and a few develop high conductivity over time in storage, although the latter cannot be relied on as a safety measure. It is also possible to add a proprietary anti-static additive at a level of about 0.15%. Small impurities of alcohols in esters or of inorganic salts can also increase conductivity by orders of magnitude. Freshly distilled water has a conductance of 5.0×10^{-8} siemen, but this rapidly increases as it picks up CO_2 from the air.

The conductivity limit that is usually regarded as safe is 1.0×10^{-10} siemen/cm (100 picosiemen/cm) and above this level it is not necessary to earth the equipment handling the solvent.

Resistivity, the reciprocal of conductivity, is also often quoted and the danger limit in various resistivity units is

100 megohm metre	(MΩ m)
$1.0 \times 10^{+4}$ megohm cm	(MΩ cm)
$1.0 \times 10^{+10}$ ohm cm	(Ω cm).

In general, all hydrocarbons and ethers (but not glycol ethers) have conductivities of 1 picosiemen/cm or less and are liable to generate static electricity. The higher molecular weight esters are at or near the limit. The unit used in the tables is siemen/cm.

The minimum ignition energy of the spark required to cause an ignition for most solvents lies in the range of 0.2–1.5 mJ, but carbon disulphide which has a very low conductivity (1.0×10^{-16} siemen/cm) also has a very low minimum ignition energy (about 0.015 mJ) and a very wide range between LEL and UEL. It thus represents an exceptionally high electrostatic hazard.

Immediate danger to life and health (IDLH)

The IDLH value represents a maximum vapour concentration from which a person can escape within 30 min without irreversible health damage or effects that would impair the ability to escape. Such information is clearly important in rescues and emergencies. It should be compared with the LEL and the saturated vapour concentration at the ambient temperature. Since a spark might cause an explosion in an atmosphere within the flammable range even if the IDLH is greater than the LEL other considerations than the IDLH may prohibit entering a solvent-laden atmosphere.

Occupational exposure standard (OES)

An OES is the exposure to a solvent in air at which there is no indication that injury is caused to people, even if it takes place on a day-after-day basis.

The long-term exposure limit to solvent vapours sets a limit for the average exposure over an 8-h working day. It applies to workers in a plant and not to people living in the neighbourhood.

The short-term exposure limit (STEL) also applies to some solvents and refers to an average over a peak period of 15 min. This is meant for the type of exposure that occurs when cleaning a filter press or doing other regular, but short-term tasks. The average over the peak would be counted as part of the 8-h exposure.

The limits vary from country to country and are constantly being reviewed in the light of experience. The figures quoted in this book are those applicable in the U.K. in 1996 and are expressed in ppm. Where a British figure is not available U.S. TLV-TWA figures are used.

Odour threshold

This is extremely subjective and hard to define accurately. In one reported test 10% of those taking part could detect an odour at 1 ppm while 50% could do so at 25 ppm. At 500 ppm there was still 10% of those exposed who could not detect it.

There is further a difference between identifying a smell and just detecting it so that complaints of an odour are hard to refute reliably and smell cannot be relied upon as a warning of potentially dangerous exposure. This is particularly true in the case of long periods of exposure since the nose becomes desensitized.

The figures quoted here are for concentrations where all the people exposed could detect, although not identify, an odour.

Solvents are not, as a class, very odiferous materials and few can be detected at much below a 1 ppm level unlike mercaptans (which can be smelt at the low ppb level), sulphides and aldehydes. The latter are often detectable in solvents that have been recovered and recycled and make such recovered solvents unacceptable for use in household formulations.

Some solvents, such as dimethylformamide (DMF), have very low odours themselves but contain trace quantities of impurity (dimethylamine in the case of DMF) which are much easier to detect. Others, e.g. dimethylsulfoxide, produce very unpleasant smells when they are degraded biologically so that even small quantities getting into an aqueous effluent are unacceptable.

Saturated vapour concentration (SVC)

The concentration of vapour in equilibrium with liquid (or solid) solvent is important for a number of reasons:

1. Fire and explosion.
2. Toxicity.
3. Smell.
4. Loss in handling.

Vapour concentration can be expressed in mgm/m^3, ppm or %. The former lends itself to ventilation calculations where the quantity of solvent being evaporated into a body of air is known.

Both ppm and percentage figures are based on volumes of solvent vapour in air and the conversion is given by

$$\text{ppm} = \text{mgm/m}^3 \times 24.04/\text{solvent molecular weight.}$$

All the SVC quoted here are at 21°C (equivalent to 70°F).

1. *Fire and explosion*. The concentration leading to a fire hazard is very much greater than that leading to a health hazard. It is unusual for someone exposed to a fire hazard not to be able to detect solvent odour by nose although, since all solvents are denser than air, the concentration at floor level may be very much greater than that at head height.

2. *Toxicity*. This is discussed elsewhere. Above the normally quoted health levels asphyxiation can take place at an SVC of about 150,000 ppm. A high concentration of inert gas (or CO_2) used for blanketing the vapour space in a tank can also be dangerous in this way.

3. *Smell*. This is discussed in the section on *Odour threshold*.

4. *Loss in handling*. Every time a bulk liquid is transferred between road tanker and storage tank or between storage and process there is a potential discharge of vapour. In addition, solvent vapours are discharged when the storage tank 'breathes' with the daily change of temperature.

Increasingly it is becoming unacceptable that this discharge goes directly into the atmosphere and the alternatives are to return the vapour to the vapour space of the vessel from which the liquid comes or to pass the solvent-rich ventings to recovery or destruction. The linking of vents and recovery can become very complicated if more than one solvent is involved in the system and destruction of the solvent in the ventings before their discharge to atmosphere is the most common solution. The loss of solvent is no greater than it would be if the ventings were discharged directly but, to design a destruction plant, the amount of discharge must be known. The most volatile solvents (pentane, ether, dichloromethane) can lose 0.3% of the liquid transferred on each occasion and in a good recovery system the handling loss can be the largest contribution to the total losses of solvent.

Vapour density relative to air

This is the ratio between the molecular weight of the solvent and the molecular weight of air. Apart from methanol which has the lowest vapour density, all organic solvents are much heavier than air.

This means that spillages, whether on a small scale in the laboratory or on a large scale in an industrial plant, will give rise to vapour at a low level. Ventilation should therefore be designed to draw from this level and tests for flammable or toxic concentrations should be made at a low point.

Heavy vapours can spread for long distances in ditches, pipe tracks and drainage pipes and can accumulate in bunded areas, particularly if the bund walls are high. The manual clearing of sludges and deposits in the bottoms of storage tanks which have contained low-flash point solvents is particularly hazardous if low-level ventilation is not provided.

Photochemical ozone creation potential (POCP)

POCP is an arbitrary scale of atmospheric chemical activity based on ethylene at 100 and the very stable organics at 0. The 'natural' products such as alpha-pinene and dipentene have a POCP of about 50.

A significantly large contribution to the total of volatile organic compounds (VOCs) in industrial countries is derived from the use of solvents. Since VOCs are an essential ingredient of smog both legislation and public opinion will lead to the choice of solvents which have a low POCP.

This is particularly true for paints and for domestic uses where recapture and recovery of the used solvent or its destruction before discharge are impractical. Since there is little correlation between the toxicity, evaporation rate, solvent power and POCP of solvents this entails a further independent restriction to the choice of solvent for domestic purposes.

The POCP should not be confused with the ozone depletion potential (ODP) which depends on the extreme stability of various halogenated solvents in the atmosphere, but, because POCP is a measure of reactivity in the complex chemistry of the lowest level of the atmosphere, solvents with a high ODP (see Table 2) do have a very low POCP.

Table 2

	POCP	*ODP*
CFC113		0.80
Methylene chloride	0.9	<0.05
1,1,1-Trichloroethane	0.1	0.15
Chloroform	1.0	
Perchloroethylene	0.5	0
Carbon tetrachloride		1.04
Trichloroethylene	6.6	0

The class of solvents with particularly high POCP is made up off aromatic hydrocarbons with methyl sidechains such as trimethyl benzenes and the xylenes. Legislation has restricted their use in Los Angeles for many years and their widespread use in paint formulations is steadily being reduced. Developments in the resins used in paints will reduce the proportion of solvents and demand increased use of more sophisticated solvents and of water in their place. If such improvements cannot replace, say, xylenes, careful fractionation can, at a price, reduce the POCP as Table 3 shows. Since *m*-xylene is usually the most common isomer in solvent C_8 aromatics, the improvement may be considerable.

Table 3

	POCP
Ethylbenzene	59.3
o-xylene	66.6
p-xylene	88.8
m-xylene	99.3

Methyl sidechains on paraffins or naphthenes do not have the same harmful effect. The replacement of acetone (POCP 17.8) by methyl acetate (POCP 2.5) or the use of isobutyl acetate (POCP 33.2) for MIBK (POCP 63.3) is typical of what may be achieved in reducing the adverse impact of solvents on the environment.

Miscibility with water

All solvents are at least partially miscible with water and most of those with a polarity of more than 36 (on a scale of water = 100) are wholly so. Moisture levels as low as 200 ppm can easily be measured by the Karl Fischer method. Only solvents with a very low solubility in water and densities of less than 1.00 should be tested by the Dean and Stark method.

The requirements for dryness in a solvent range from the low ppm for a Grignard reagent solvent to 2 or 3% for cellulose paint thinners or gun washes. While most can be dried by various forms of distillation, there are also many solid dessicants using chemisorption or hydration effects, although none of these dessicants are general purpose. Molecular sieves are very effective and are suitable to dry the great majority of solvents. However, unless regeneration plant is used to recover the molecular sieve, their cost is about £10,000/tonne of water removed.

Many solvents are hygroscopic and if moisture is to be kept at a very low level, the vents of storage tanks should be fitted with silica gel or molecular sieve-filled canisters. For the removal of small amounts of solvents from water see the section on log activated carbon partition.

Knowledge of the solubility of solvents in water is useful in predicting their behaviour in several fields. Highly water-soluble solvents carry materials and migrate themselves into the biosphere. They are both more easily leached from soil and less easily volatilized into air. The large number of solvents that are not fully miscible with water at 25°C is a measure of the high difference between the polarity of water and that of many organic solvents. Table 4 lists a number of other solvent pairs that have an upper critical solution temperature at a temperature within normal industrial operating range.

The polarity of the non-polar solvents are all less than 6.5 (on the scale of water = 100), while the polar solvents have a polarity of 30 or more.

Log_{10} activated carbon partition

While aqueous effluents containing highly volatile solvents can be stripped using air or steam preparatory to being discharged, the less volatile and particularly those that are polar, are difficult to strip and are more economically removed from dilute solution using activated carbon- or ion-exchange resins.

To get an idea of the effectiveness of activated carbon adsorption, one can use the following equation as a preliminary guide, although an experiment using the grade of carbon to be used is vital to get a sound design

Table 4 Upper critical solution temperature (°C)

	nC_5	nC_6	nC_7	nC_8	nC_9	nC_{10}	C_6	CS_2	*2,2,4-TMP*
Methanol	14.8	35	51	67		76	45	36	42.5
Ethanol	<–78	–65	–60			–15	–16	–24	–70
EGME		28	49					25	40
EEE		–32	–12				–60		–15
Carbitol		12	25					<–1	28
Acetone		–39	–28	–6		–6	–40	–29	–34
Acetophenone		3	4			10		–16	14
DMF	63	68	73					50	
Acetic acid		–4	–8	19	29	41	3.9	7	7
Aniline	72	69	70	72	75	78		30	80
Nitrobenzene	25	20	18	20	22	24		–4	29
Pyridine		–25	–22					–36	–15
Acetonitrile	60	77	85	92	100	108		77	81
Furfural		92	94					66	101
Phenol	57	51	60						

$$P_{AC} = \frac{x}{m} \cdot \frac{1}{c}$$

x/m is the weight of adsorbate in mg per kg of charcoal, c is the concentration in ppb of the solute remaining in the effluent, and P_{AC} is the activated charcoal partition coefficient.

The values of P_{AC} are usually quoted in logarithms to the base 10. For a rough preliminary estimate

$$\log P_{AC} = 6 - \log S$$

where S is the solubility of the solvent in water in ppb. This relationship is not valid if the solute is wholly water miscible, and clearly shows that the more water-soluble a solvent is the less easy it is to remove it from water by adsorption.

The partition coefficient is affected by temperature, pH and the type of activated charcoal used.

Log_{10} partition between octanol and water

A great deal of work on the partition of solutes between water and other solvents has been done by Pomona College. The main solvent used is *n*-octanol giving the relationship

$$P_{ow} = \frac{\text{concentration of solute in } n\text{-octanol}}{\text{concentration of solute in water}}$$

Originally the work was done as a guide to the biological effect of the solute. A high value of P (i.e. $\log P > 1.5$), corresponding to a low concentration in water, means that the material in solution cannot easily invade living organisms and therefore has a low biological effect. On the other hand, a negative value of $\log P$ indicates a very hydrophilic compound, difficult to extract from water using any third solvent, not just *n*-octanol.

The values of P_{ow} are expressed as logarithms to base 10 and the logarithms have a range of −1.93 to +4.15, a range of a million.

A reasonable estimate for P, if experimental results are not available, is

$$\log P_{ow} = 6.75 - 0.75 \log S$$

where S is the solubility of the solute in water in ppb.

Oxygen demand

The biodegradability of solvents to the simplest molecules, primarily CO_2 and water in a given time (here the quoted biological oxygen demand figure is for five days except for a few instances where 10 days are quoted) vary widely and the correlation between laboratory and plant-scale results for the amount of oxygen removed from the aqueous phase is not very reliable.

The theoretical oxygen demand (ThOD), is solely the oxygen needed on a stochiometric basis to oxidize the solvent completely, and is thus the worst possible effect, but may be useful if no laboratory results are available. In this book the values of ThOD do not include for the oxidation of the nitrogen where it exists in the solvent's molecule. This tends to be a slow reaction and seldom is represented in the five-day BOD test.

BOD depends on the effectiveness of the organisms that may be present and which may be killed by a change of the solvent in the effluent and starved to death by a lack of the solvent to which it is accustomed. Results for BOD can be measured over a time period measured in days, usually five or 10, and is clearly a time-consuming test.

A high BOD solvent sparingly miscible in water and with no solvent-rich phase present to replenish the aqueous phase may be less harmful than a low BOD solvent that is readily soluble in water.

Antoine vapour pressure equation

There is one very widely used equation for estimating the vapour pressure of organic liquids, the Antoine equation

$$\log P = A - \frac{B}{C + T}$$

where A, B and C are constants.

P is the vapour pressure of the solvent at temperature *T* which can be expressed in a number of pressure units which, of course, refer to different values of *A*. It is therefore most important to know what pressure units are in use when getting values of the constants from the literature.

In this book logarithms to base 10, mmHg and °C are used.

Cox chart equation

As an alternative to using the Antoine equation it is possible to employ an equation based on the Cox chart

$$\log P = A - \frac{B}{T + 230}$$

These constants are not the same as the Antoine ones although they tend to coincide when $C = 230$. In this book the same units for *P* and *T* are used and the logarithms are also to base 10. Since another correlation gives a value for *C* of

$$C = 239 - 0.19T_b$$

A and *B* from the two systems will be close together for solvents boiling near 50°C.

Since both Cox and Antoine equations are based on the Clausius–Claypeyron equation the values for *B* are related to the latent heat of the solvent.

The Cox equation lends itself to calculating relative volatilities

$$\log \alpha^* = \log \frac{P_1}{P_2} = A_1 - A_2 - \frac{B_1 - B_2}{230 + T}$$

and the sensitivity of the value of the relative volatility to temperature tends to be high when the difference in latent heats is high. Alcohols tend to have higher molal latent heats than other groups of solvents and therefore to have large changes in α^* with changes of temperature and pressure.

Solubility parameter (δ)

In choosing a solvent for a particular duty, knowledge of its solubility parameter can be of considerable assistance. A resin, a polymer or any other non-electrolyte is likely to be most easily soluble in a solvent if the solubility parameters of the solvent and the solute are similar. It follows that two solvents with similar parameters will have similar dissolving powers for a given resin

$$\delta = \left(\frac{L - RT}{V}\right)^{1/2}$$

where *L* is the molal latent heat of the solvent, *T* the absolute temperature and *V* its molar volume. The value of is normally expressed in units of $cal^{1/2}$ $cm^{-3/2}$.

The solubility parameter is a blendable property so its value for a blend can be calculated and a blend of solvents can be made to suit the solute more closely if necessary. While the value of δ can be calculated knowing only readily available information, it can also be found experimentally for hydrocarbons and chlorinated hydrocarbons by measuring the Kauri Butanol number (*KB*) since

$$KB = 50\delta - 345 \text{ (for } KB > 35).$$

Just as a solute with a similar value of δ to the solvent will dissolve so two solvents of similar δ will be miscible. The limit of difference beyond which total miscibility will not be achieved at 298 K is about 2.5, but as the values of UCST show (*Miscibility with water* section), at higher temperature miscibility becomes easier.

Dipole moment

The figures quoted here are for liquids at 298 K. The dipole moment is proportional to $1/T$ where *T* is the absolute temperature.

Along with a number of other properties, dipole moment contributes to the 'polarity' of a solvent.

Dielectric constant

The dielectric constant of a solvent reflects its molecular symmetry and is comparatively easy to measure. It can thus be used to calculate molar polarization and, from it, dipole moment (P).

$$P = \frac{\varepsilon - 1}{\varepsilon + 2} V$$

where ε is the dielectric constant and V the molecular volume.

Dielectric constant is also a factor in considering a solvent's electrostatic hazard. A solvent's relaxation time, which is a measure of the rate at which an electrostatic charge will decay, is a product of dielectric constant and resistivity. The higher this product the higher the relaxation time.

However, the range of values of the dielectric constant is about 2–180, which is a small range compared to the range of resistivity (see *Electrical conductivity* section). Nonetheless if a solvent is being changed in an existing process the possible increased risk of electrostatic problems should not be ignored.

Polarity

Polarity is a widely discussed and quoted property of a solvent but it is used loosely to cover a number of different effects, including those covered by dielectric constant and dipole.

The figures quoted here are comprehensively discussed in reference 14 of the bibliography.

Evaporation time

There is no satisfactory method of calculating the rate of evaporation of a solvent, since it depends on the equipment in which evaporation takes place as well as a number of properties of the solvent.

There are two widely used standard solvents – diethyl ether and butyl acetate – against which other solvents' evaporation times can be compared. Somewhat confusingly, a low rate of evaporation on the ether scale corresponds to a high number (i.e. the time it takes to evaporate is many times the time ether takes), while on the butyl acetate scale a low rate of evaporation corresponds to a low number (i.e. the rate of evaporation is lower than that of butyl acetate).

An approximate relationship between the two scales is

$$B = 15/E$$

where B and E are the butyl acetate and ether numbers.

Nett heat of combustion

For the eventual disposal of used solvent, whether in liquid or vapour form, the preferred method is usually burning. This may involve using the solvent as a fuel, possibly in a cement or lime kiln in which it may be used as a replacement for a more conventional, and more costly fuel. Alternatively, to reduce VOCs to a very low level in solvent-laden air, combustion in an incinerator, with or without added fuel, may be used.

A less common alternative is oxidation in the liquid phase which also gives rise to heat.

In all cases the heat of combustion of the solvent to be destroyed needs to be known. Since in almost every case the water generated in the destruction will be discharged as vapour, the lower or nett calorific value is the appropriate one to use and it is the one quoted here.

Heat of fusion

It may be theoretically important to know how much heat will be required to thaw out the solvent should it freeze, but the most frequent use of the heat of fusion is to estimate the freezing point depression when the solvent dissolves a solute.

The freezing point depression per gram-mole in 100 grams of solvent can be adequately calculated up to a mole fraction of solute of 0.10 by the expression

$$\frac{RT^2}{100H}$$

where R is the gas constant (1.987 cal/g units), H is the latent heat of fusion in cal/g and T is the freezing point of the pure solvent in K.

For some commonly used solvents with high melting points the freezing point depression coefficients are listed below.

Acetic acid	39.0
Benzene	49.0
t-Butanol	80.6
Cyclohexane	208
Cyclohexanol	420
Dimethylsulphoxide	40.9
1,4-Dioxane	46
Nitrobenzene	70
Phenol	74
Sulfolane	641
Water	18.6

Azeotropes

The presence of an azeotrope between solvent components can have three important effects:

1. It makes the recovery by distillation of one of the solvents to a high degree of purity and a high yield difficult. Azeotropic mixtures should therefore be avoided if possible in pharmaceutical production where recovery is important.
2. It increases the rate of evaporation in the great majority of cases since the azeotrope is usually a low boiling one in which the boiling point of the azeotrope is below that of both pure components.
3. It decreases the flash point of the mixture and can therefore have an important influence on safety when the solvent mixture may be used at a temperature about ambient. It is important, therefore, to know when an azeotrope exists and also when its absence is confirmed.

In this book binary azeotropes mostly drawn from *Horsley's Azeotropic Data* are listed. There are a great number of possible ternary or more complex mixtures of solvents that are used, and some ternary azeotropes have been recorded. It is extremely rare for a ternary azeotrope to occur if all three binary mixtures which its components can form are not also azeotropic.

In the absence of information on the existence of an azeotrope in a binary mixture of solvents it is possible to estimate whether an azeotrope will exist if activity coefficients at infinite dilution (γ^∞) and pure vapour pressures of the solvents (P) are available.

In an ideal vapour–liquid system of two solvents the relative volatility (α^*) is equal to the ratio of the vapour pressure of the components and is not affected by composition

$$P_1/P_2 = \alpha^*$$

An azeotrope occurs in a non-ideal system when the vapour phase and the liquid phase in equilibrium with it have the same composition

$$\alpha = \frac{\gamma_1 P_1}{\gamma_2 P_2} = \frac{\gamma_1}{\gamma_2}\alpha^* = 1.0$$

The values of γ vary throughout the concentration range. By definition $\gamma_1 = 1.0$ for pure component 1 at the composition at which component 2 is infinitely dilute and has an activity coefficient γ_2^∞ and vice versa.

A low-boiling azeotrope (much the more common) will occur if $\gamma_2^\infty > \alpha^*$, while a high-boiling azeotrope will occur if $1/\alpha^* > \gamma_1^\infty$.

Activity coefficients (γ)

Activity coefficients are described as a measure of the relative escaping tendency of a compound. This may be escape from a liquid phase to a vapour phase (which can also be quantified by Henry's law coefficient) or from one liquid phase to another, which is the basis of liquid–liquid extraction.

The data sheets list the experimental values of γ^∞ which are obtained by a variety of methods. These can show an appreciable difference one from another, particularly when the mixtures they refer to are very non-ideal with values of γ^∞ of 10^3 or more. There is also a considerable temperature effect on values of γ.

The two major sources of the values listed here are references 2a and 2c. Those from 2a, which have reference numbers 1/1 to 8/381 are derived from distillation data and are at temperatures near boiling point, although these may be at pressures lower than atmospheric for high boiling or unstable solvents. Those from 2c are measured at 25°C and have reference numbers commencing 1X. The latter are more useful for liquid–liquid extraction calculations.

If no experimental results are available for particular solute/solvent mixtures UNIFAC provides a method of calculating γ^∞ values.

Partition coefficient (K)

There are an almost unlimited number of systems which can involve solvent extraction, but those listed here are restricted to the ones which have water as phase 1, an extraction solvent, sparingly soluble with water, as phase 2 and a solute partitioned between the two phases.

$$K = \frac{\text{molar concentration of solute in water}}{\text{molar concentration of solute in X}}$$

If the objective is to remove the solute from phase 2 and concentrate it in phase 1, K needs to be as high as possible and a target minimum of 5.0 is appropriate. On the other hand, if the solute needs to be removed from the water phase, K should be 0.2 or less.

The values of partition coefficient listed are for 25°C and the most dilute solutions for which information was available. K is temperature dependent and tends to 1.0 as the concentrations of solute in the two phases increase. Since the two phases each containing solute are in equilibrium, the product of the mole fraction of solute and its activity coefficient in each phase is the same. Hence

$$K = \gamma_2/\gamma_1$$

For dilute solutions the activity coefficients can be used to assess the likely performance of the system. Thus for the system water (1)–heptane (2)–MEK (solute)

$$K = \frac{\gamma_2^\infty}{\gamma_1^\infty} = \frac{3.91}{27.2} = 0.14$$

This would indicate that it was possible to remove MEK from water using heptane as an extraction solvent.

The references which give the source of the information are from three sources in the bibliography. These are:

1. CEH: *Perry's Chemical Engineering Handbook*
2b. V2, 3 and 4: *Dechema Liquid/liquid Data Collection*
17 P: *Pomona College Collection*.

Henry's law constant (H)

Particularly in dilute solutions in water, solvents tend to behave in a very non-ideal way and their equilibrium vapour pressure has to be calculated either using an activity coefficient or Henry's law constant (H). The literature contains compilations of the latter for aqueous solutions but they are reported in several different units, all of which are a pressure divided by a concentration, i.e. $H = P/x$, where P is the vapour pressure of the pure solvent at the solution temperature and x its concentration in the liquid phase.

In this book H is expressed in atmospheres divided by mole fractions. Alternative units are

1. Atmospheres per g-mole of solvent per 100 m^3 of water. Convert by multiplying by $10^6/18$.
2. Kilopascals per g-mole of solvent per 100 m^3 of water. Convert by multiplying by 548.
3. Atmospheres per lb-mole per ft^3. Convert by multiplying by 6.25×10^{-5}.

The value of H increases with temperature and the figures here are for the system temperature of 25°C.

Figures for H quoted in the literature for apparently identical systems vary widely, sometimes by an order of magnitude or more, but, if the information is available there are two ways of checking it.

1. Since H is only suitable for use in dilute solutions

$$H = P\gamma^\infty$$

If therefore figures for the activity coefficient at infinite dilution and 25°C and the Antoine coefficients are available, the value of H can be compared.

2. Many solvents, particularly hydrocarbons, chlorinated, and the higher molecular weight oxygenated ones, are so insoluble in water that their aqueous solutions are always dilute. At saturation, therefore

$$H = P/S$$

where S is the solubility of the solvent in water expressed as a mole fraction.

High values of H (e.g. > 50) indicate a dis solved solvent that can be stripped easily either by air or steam. Such a solvent will also evaporate quickly from water.

H can also be used to calculate the composition of solvent-laden air in contact with water at levels appropriate to TLV calculations thus

$$C = \frac{(\text{TLV in ppm}) \cdot (\text{mol. wt of solvent})}{H \cdot 18}$$

where C is the concentration of solvent in water which corresponds to the TLV.

Similarly the flash point of a dilute aqueous solution can be seen to be above or below 25°C given a value for the LEL of the pure solvent.

Section 1

Hydrocarbons

n-Pentane

Alternative names
Below 40°C, petroleum ether

Reference codes

CAS number	109 66 0	Hazchem code	3Y
UN number	1265	EPA code	

Physical properties

Molecular weight	72	Cubic expansion coeff (per °C × 10^3)	1.52
Empirical formula	C_5H_{12}	Surface tension (@20°C dyn/cm)	16
Boiling point (°C)	36	Absolute viscosity (@25°C cP)	0.235
Freezing point (°C)	−129	Refractive index (20°C)	1.358
Specific gravity (20/4)	0.626		

Fire hazards

Flash point (closed cup °C)	−40	Lower explosive limit (ppm)	15000
Autoignition temperature (°C)	260	Upper explosive limit (ppm)	78000
Electrical conductivity	2E−10		

Health hazards

IDLH (ppm)	5000	Vapour concentration @21°C ppm	768000
OES-TWA	600	Vapour density (relative to air)	2.5
OES-STEL	750	Vapour pressure @21°C mmHg	442
Odour threshold (ppm)	900	POCP	41

Aqueous effluent

Solubility in water (25°C %w/w)	38E−4
Solubility of water in (25°C %w/w)	120E−4
Log_{10} activated carbon partition	
Log_{10} partition in octanol/water (w/w)	+3.23
Biological oxygen demand w/w (days)	
Theoretical oxygen demand w/w	3.56

Vapour pressure equation constants (Log_{10}, mmHg)

Antoine equation	A	6.87632
	B	1075.780
	C	233.205
Cox chart	A	6.82847
	B	1050.1

Solvent properties

Solubility parameter	7.0	Kauri butanol value	28
Dipole (D)	0	Evaporation time (ether = 1)	1.0
Dielectric constant (20°C)	1.844	Evaporation time (BuAc = 1)	13
Polarity (water 100)	0.9		

Thermal information

Latent heat (cal/mol)	6120
Nett heat of combustion (kcal/gmol)	776
Specific heat (cal/mol/°C)	40.32
Critical pressure (MN/m^2)	3.31
Critical temperature (K)	470
Latent heat of fusion (cal/mol)	2008
Van der Waals' volume	3.825
Van der Waals' surface area	3.316
Molar volume	115.0

Solute	Azeotrope X% w/w	°C	Reference	Solute γ^∞	Reference	Partition coefficient	Reference
Hydrocarbons							
n-Pentane	–		–	–	–	–	–
n-Hexane	None			0.9	6a/123		
n-Heptane	None			0.8	6a/127		
n-Octane				1.2	1x3/1149		
n-Nonane							
n-Decane							
2,2,4-TMP							
Cyclohexane	None		9742	1.1	6a/119		
Benzene	None		9741	1.6	6a118		
Toluene	None			3.5	6c/160		
Ethylbenzene							
Xylenes							
C_9 Aromatics							
Tetralin							
Alcohols							
Methanol	8	31	2055	19.0	2e/132		
Ethanol	95	34	4062	13.6	2c/375		
n-Propanol	None		6484				
i-Propanol	94	35	6370				
n-Butanol	None			8.5	2b/169		
i-Butanol	None		8327				
s-Butanol	None		8228				
n-Amyl alc.							
i-Amyl alc.							
Cyclohexanol							
1-Octanol	None			8.7	2f/529		
Ethanediol	None						
DEG							
1,2-Propanediol							
Glycol ethers							
PGME							
EGME							
EEE							
EGBE							
Chlorinated							
MDC	51	3	1571	2.4	6a/100		
Chloroform	None		1482				
Carbon tet.							
1,2-EDC							
1,1,1-TCA							
TCE							
Perk.							
MCB							

Hydrocarbons

Solute	Azeotrope X% %w/w	Azeotrope °C	Azeotrope Reference	Solute γ^∞	Reference	Partition coefficient	Reference
Ketones							
Acetone	80	32	5368	7.1	3+4/190	0.91	V2/475
MEK				5.3	1x/3/1149		
MIBK							
Cyclohexanone							
NMP							
Acetophenone							
Ethers							
Diethyl ether	32	33	8296				
DIPE							
Dibutyl ether							
MTBE							
1,4-Dioxane				4.8	1x/3/1149		
THF	None			2.2	1x/3/1149		
Esters							
Me acetate	78	34	5536				
Et acetate							
i-Propyl acetate							
n-Butyl acetate							
Cellosolve acetate							
Miscellaneous							
DMF							
DMAc							
DMSO							
Sulfolane							
CS_2	89	36	1256				
Acetic acid	None		3156				
Aniline							
Nitrobenzene	None		9740				
Morpholine							
Pyridine							
2-Nitropropane							
Acetonitrile	90	35	2792	22.0	6a/102		
Furfuraldehyde							
Phenol							
Water	99	35	462				

n-Hexane

Alternative names
62/68 Hexane

Reference codes

CAS number	110 54 3	Hazchem code	3YE
UN number	1208	EPA code	

Physical properties

Molecular weight	86	Cubic expansion coeff (per °C × 10^3)	1.3
Empirical formula	C_6H_{14}	Surface tension (@20°C dyn/cm)	18.4
Boiling point (°C)	69	Absolute viscosity (@25°C cP)	0.31
Freezing point (°C)	–95	Refractive index (25°C)	1.372
Specific gravity (20/4)	0.659		

Fire hazards

Flash point (closed cup °C)	–22	Lower explosive limit (ppm)	12000
Autoignition temperature (°C)	225	Upper explosive limit (ppm)	75000
Electrical conductivity	1.0E–16		

Health hazards

IDLH (ppm)	5000	Vapour concentration @21°C ppm	170000
OES-TWA	20	Vapour density (relative to air)	2.99
OES-STEL		Vapour pressure @21°C mmHg	128
Odour threshold (ppm)		POCP	42

Aqueous effluent

Solubility in water (25°C %w/w)	9.5E–4
Solubility of water in (25°C %w/w)	110E–4
Log_{10} activated carbon partition	
Log_{10} partition in octanol/water (w/w)	+3.80
Biological oxygen demand w/w (days)	0.04 (7)
Theoretical oxygen demand w/w	3.53

Vapour pressure equation constants (Log_{10}, mmHg)

Antoine equation	A	6.91058
	B	1189.64
	C	226.28
Cox chart	A	6.9386
	B	1212.1

Solvent properties

Solubility parameter	6.9	Kauri butanol value	30
Dipole (D)	0	Evaporation time (ether = 1)	1.4
Dielectric constant (20°C)	1.9	Evaporation time (BuAc = 1)	8.4
Polarity (water 100)	0.9		

Thermal information

Latent heat (cal/mol)	6880
Nett heat of combustion (kcal/gmol)	921
Specific heat (cal/mol/°C)	42.0
Critical pressure (MN/m^2)	3.03
Critical temperature (K)	507.5
Latent heat of fusion (cal/mol)	3119
Van der Waals' volume	4.50
Van der Waals' surface area	3.86
Molar volume	130.5

Solute	Azeotrope X% w/w	°C	Reference	Solute γ^{∞}	Reference	Partition coefficient	Reference
Hydrocarbons							
n-Pentane	None			0.9	6a/123		
n-Hexane	–		–	–	–	–	
n-Heptane	None		12133	0.9	6a/604		
n-Octane	None		12134	0.5	6a/613		
n-Nonane							
n-Decane							
2,2,4-TMP							
Cyclohexane	None		11690	1.1	6a/273		
Benzene			10861	1.4	6a/535		
Toluene	None		12131	1.4	6a/591		
Ethylbenzene	None						
Xylenes	None			1.5	6a/605		
C_9 Aromatics							
Tetralin							
Alcohols							
Methanol	72	50	2087	39.1	2a/253		
Ethanol	79	59	4106	12.3	2a/453	38.1	P383
n-Propanol	96	66	6506	22.4	2a/584	0.74	V2/577
i-Propanol	77	63	6390	10.5	2b/97	1.00	V2/616
n-Butanol	97	68	8163	11.2	2b/200	0.04	V3/121
i-Butanol	98	68	8354	13	2b/320		
s-Butanol	92	67	8242	9.5	2b/250		
n-Amyl alc.	None		9758				
i-Amyl alc.	None		9845				
Cyclohexanol	None						
1-Octanol				6.0	2f/532		
Ethanediol							
DEG							
1,2-Propanediol							
Glycol ethers							
PGME							
EGME	None						
EEE	95	66	8442	575.0	2b/295		
EGBE							
Chlorinated							
MDC	None		1575				
Chloroform	16	60	1495	1.3	6a/426		
Carbon tet.	None			1.2	6a/403		
1,2-EDC	None						
1,1,1-TCA	71	67	2730b	1.3	6a/473		
TCE	None		2330	1.5	6a/463		
Perk.	None		2217a	1.4	6a/453		
MCB	None		10519	1.8	6a/529		

Hydrocarbons

Solute	Azeotrope X% w/w	Azeotrope °C	Azeotrope Reference	Solute γ^{∞}	Reference	Partition coefficient	Reference
Ketones							
Acetone	41	50	5385	4.2	3+4/225	0.8	V2/487
MEK	71	64	7376	3.5	3+4/302	0.13	V3/26
MIBK							
Cyclohexanone							
NMP							
Acetophenone				17.6	3b/451		
Ethers							
Diethyl ether	None						
DIPE	47	67	12128				
Dibutyl ether							
MTBE				1.2	1x/3/1240		
1,4-Dioxane	98	60		3.0	3+4/472		
THF	50	63		1.6	1x/3/1240	0.16	
Esters							
Me acetate	39	52					
Et acetate	62	65	7588	2.6	5/514		
i-Propyl acetate	91	69	9297				
n-Butyl acetate							
Cellosolve acetate							
Miscellaneous							
DMF	None				6c/332	24.8	V2/547
DMAc	None						
DMSO							
Sulfolane				121	1x/3/1240		
CS_2	None		1274				
Acetic acid	95	68	3184	13.4	5/152	9.2	V2/282
Aniline	None		11151	11.6	6a/580		
Nitrobenzene	None		10708	14.5	6a/532		
Morpholine							
Pyridine	None						
2-Nitropropane	97	69	6284	7.4	6a/510		
Acetonitrile	72	52	2800				
Furfuraldehyde							
Phenol						1.91	P1659
Water	94	62	570				

n-Heptane

Alternative names

Reference codes

CAS number	142 82 5	Hazchem code	3YE
UN number	1206	EPA code	

Physical properties

Molecular weight	100	Cubic expansion coeff (per °C × 10^3)	1.3
Empirical formula	C_7H_{16}	Surface tension (@20°C dyn/cm)	19.3
Boiling point (°C)	98	Absolute viscosity (@25°C cP)	0.41
Freezing point (°C)	–91	Refractive Index (25°C)	1.385
Specific gravity (20/4)	0.664		

Fire hazards

Flash point (closed cup °C)	–4	Lower explosive limit (ppm)	10000
Autoignition temperature (°C)	215	Upper explosive limit (ppm)	70000
Electrical conductivity	1.0E–16		

Health hazards

IDLH (ppm)	4250	Vapour concentration @21°C ppm	55000
OES-TWA	400	Vapour density (relative to air)	3.57
OES-STEL	500	Vapour pressure @21°C mmHg	40
Odour threshold (ppm)	150	POCP	53

Aqueous effluent

Solubility in water (25°C %w/w)	3E–4
Solubility of water in (25°C %w/w)	100E–4
Log_{10} activated carbon partition	
Log_{10} partition in octanol/water (w/w)	+3.5
Biological oxygen demand w/w (days)	0 (7)
Theoretical oxygen demand w/w	3.52

Vapour pressure equation constants (Log_{10}, mmHg)

Antoine equation	A	6.89386
	B	1264.370
	C	216.64
Cox chart	A	7.04265
	B	1365.1

Solvent properties

Solubility parameter	7.5	Kauri butanol value	31
Dipole (D)	0.0	Evaporation time (ether = 1)	3.0
Dielectric constant (20°C)	1.924	Evaporation time (BuAc = 1)	3.3
Polarity (water 100)	1.2		

Thermal information

Latent heat (cal/mol)	7560
Nett heat of combustion (kcal/gmol)	1067
Specific heat (cal/mol/°C)	52
Critical pressure (MN/m^2)	2.74
Critical temperature (K)	540
Latent heat of fusion (cal/mol)	3378
Van der Waals' volume	5.17
Van der Waals' surface area	4.40
Molar volume	147.5

Solute	Azeotrope X% w/w	°C	Reference	Solute γ^∞	Reference	Partition coefficient	Reference
Hydrocarbons							
n-Pentane				1.3	6a/127		
n-Hexane	None		12133	0.9	6a/604		
n-Heptane	–		–	–	–	–	–
n-Octane				1.0	6b/196		
n-Nonane							
n-Decane							
2,2,4-TMP	None		13809	1.0	6b/197		
Cyclohexane	None		11697	1.1	6a/304		
Benzene	0.7	80	10876	1.3	6b/123	<0.001	P1487
Toluene	None		13027	1.4	6b/169	0.003	P2290
Ethylbenzene	None		13807	1.7	6c/489	<0.001	P2877
Xylenes	None		13808	1.4	6c/497	<0.001	P2877
C_9 Aromatics				1.3	6b/200		
Tetralin							
Alcohols							
Methanol	59	49	2101	18.4	2c/243		
Ethanol	51	71	4139	11.6	2a/498	5.41	V2/376
n-Propanol	64	88	6514	15.5	2a/596	0.46	V2/583
i-Propanol	50	76	6399	14.5	2b/113	1.18	V2/620
n-Butanol	82	94	8182	7.6	2b/218		
i-Butanol	73	91	8368		2d/378		
s-Butanol	62	89	8248	8.0	2d/281		
n-Amyl alc.	Azeo			19.0	2f/382		
i-Amyl alc.	93	98	9869				
Cyclohexanol	None		11727	23.7	2f/419		
1-Octanol				8.2	2f/535		
Ethanediol	97	98	4312				
DEG							
1,2-Propanediol							
Glycol ethers							
PGME							
EGME	77	92	6592	10.0	1x/1/359		
EEE	86	97	8461			0.90	P3982
EGBE							
Chlorinated							
MDC				2.2	1x/1/357		
Chloroform				1.3	6b/77		
Carbon tet.	None		1167	1.2	6b/67		
1,2-EDC	24	81	3009	2.3	6c/444		
1,1,1-TCA							
TCE	None		2335	1.3	1x/3/1328		
Perk.							
MCB	None		10531	1.7	6b/119		

Solute	Azeotrope X% w/w	Azeotrope °C	Azeotrope Reference	Solute γ^∞	Reference	Partition coefficient	Reference
Ketones							
Acetone	10	56	6393	3.1	3+4/242	0.66	CEH
MEK	30	77	7384	3.9	3+4/311	0.13	V3/29
MIBK	87	98	11801	2.5	3b/550		
Cyclohexanone				3.8	3b/509		
NMP				10.2	3b/460		
Acetophenone							
Ethers							
Diethyl ether				1.2	1x/3/1330		
DIPE	None		12196	4.2	3+4/559		
Dibutyl ether				1.0	3+4/592		
MTBE				1.2	1x/3/1331		
1,4-Dioxane	56	92	7552	2.7	3+4/478		
THF				1.4	1x/3/1330	0.08	V4/240
Esters							
Me acetate	3	57	5558	4.2	1x/1/358		
Et acetate	6	77	7594	3.5	1x/1/359		
i-Propyl acetate	33	88	9302				
n-Butyl acetate	None		11826a	2.0	5/591		
Cellosolve acetate							
Miscellaneous							
DMF	95	97		17.0	6b/98		
DMAc							
DMSO							
Sulfolane				108	1x/3/1330		
CS_2	None		1278	1.3	1x/1/357		
Acetic acid				18.1	5/175	10.4	V4/201
Aniline	None		11179	13.1	6b/161	0.20	P1722
Nitrobenzene							
Morpholine							
Pyridine	75	96	8860	3.1	6b/116		
2-Nitropropane	79	95	6289	5.0	6b/100		
Acetonitrile	54	69	2803	25.7	6b/79		
Furfuraldehyde	95	98	8781	10.6	3+4/50		
Phenol	None		10936	20.0	2f/400	1.3	P1657
Water	87	79	654				

n-Octane

Alternative names

Reference codes

CAS number	111 65 9	Hazchem code	3YE
UN number	1262	EPA code	

Physical properties

Molecular weight	114	Cubic expansion coeff (per °C × 10^3)	1.2
Empirical formula	C_8H_{18}	Surface tension (@20°C dyn/cm)	21.7
Boiling point (°C)	126	Absolute viscosity (@25°C cP)	0.50
Freezing point (°C)	–57	Refractive index (25°C)	1.395
Specific gravity (20/4)	0.703		

Fire hazards

Flash point (closed cup °C)	13.3	Lower explosive limit (ppm)	10000
Autoignition temperature (°C)	220	Upper explosive limit (ppm)	65000
Electrical conductivity			

Health hazards

IDLH (ppm)	3750	Vapour concentration @21°C ppm	15700
OES-TWA	300	Vapour density (relative to air)	4.1
OES-STEL	375	Vapour pressure @21°C mmHg	12
Odour threshold (ppm)	200	POCP	49

Aqueous effluent

Solubility in water (25°C %w/w)	0.63E–4
Solubility of water in (25°C %w/w)	80E–4
Log_{10} activated carbon partition	
Log_{10} partition in octanol/water (w/w)	+4.0
Biological oxygen demand w/w (days)	
Theoretical oxygen demand w/w	3.51

Vapour pressure equation constants (Log_{10}, mmHg)

Antoine equation	A	6.93142
	B	1358.800
	C	209.855
Cox chart	A	7.14736
	B	1518.9

Solvent properties

Solubility parameter		Kauri butanol value	
Dipole (D)		Evaporation time (ether = 1)	
Dielectric constant (20°C)		Evaporation time (BuAc = 1)	1.23
Polarity (water 100)	1.2		

Thermal information

Latent heat (cal/mol)	8265
Nett heat of combustion (kcal/gmol)	1213
Specific heat (cal/mol/°C)	59.3
Critical pressure (MN/m^2)	2.49
Critical temperature (K)	568
Latent heat of fusion (cal/mol)	4926
Van der Waals' volume	5.85
Van der Waals' surface area	4.93
Molar volume	163.5

Solute	Azeotrope X% w/w	°C	Reference	Solute γ^∞	Reference	Partition coefficient	Reference
Hydrocarbons							
n-Pentane				1.0	1x/1/395		
n-Hexane				0.9	6a/613		
n-Heptane	None		12133	1.0	6b/196		
n-Octane	–		–	–	–	–	–
n-Nonane							
n-Decane							
2,2,4-TMP	None		13809	0.9	6b/283		
Cyclohexane	None		11697	1.0	6a/323		
Benzene	0.7	80	10876	1.2	6b/242		
Toluene	None		13027	1.1	6b/261		
Ethylbenzene	88	126	14101	1.2	6b/273		
Xylenes	None		14120	1.2	6b/275		
C_9 Aromatics							
Tetralin							
Alcohols							
Methanol	37	72	2113	15.4	2c/249		
Ethanol	29	98	4139	21.9	2c/462	6.42	V2/385
n-Propanol	64	88	6514	3.7	2c/576	1.18	V2/585
i-Propanol	50	76	6399	7.0	2b/115	2.55	V2/622
n-Butanol	82	94	8182	4.8	2f/207	1.02	P960
i-Butanol	73	91	8368				
s-Butanol	89	62	8284	26.7	1x/3/1368		
n-Amyl alc.	Azeo			4.1	2f/383	0.24	P1270
i-Amyl alc.	93	98	9869				
Cyclohexanol	None		11727				
1-Octanol							
Ethanediol	97	98	4312				
DEG							
1,2-Propanediol							
Glycol ethers							
PGME	82	123	10002				
EGME	77	93	6592				
EEE	86	97	8461	4.5	2b/302		
EGBE	None			5.0	2b/432	0.93	P3982
Chlorinated							
MDC				2.2	1x/1/393		
Chloroform				1.4	1x/1/393		
Carbon tet.	None		1168a	1.0	6b/234		
1,2-EDC	24	81	3009	2.9	1x/1/393		
1,1,1-TCA							
TCE							
Perk.	8	121	2227				
MCB	None		10531				

Hydrocarbons

Solute	Azeotrope			Solute γ^∞	Reference	Partition coefficient	Reference
	X% w/w	°C	Reference				
Ketones							
Acetone	10	56	5393	6.1	3b/224	0.30	V2/507
MEK	30	77	7384	3.3	3+4/317		
MIBK	87	98	11801	2.1	1x/1/395		
Cyclohexanone							
NMP				14.4	1x/3/1368		
Acetophenone							
Ethers							
Diethyl ether							
DIPE	None		12196				
Dibutyl ether							
MTBE							
1,4-Dioxane	56	92	7552	2.5	3+4/480		
THF				1.5	1x/1/394		
Esters							
Me acetate	3	57	5558				
Et acetate	6	77	7594	2.4	1x/1/394		
i-Propyl acetate	33	88	9302				
n-Butyl acetate	None		11826a				
Cellosolve acetate							
Miscellaneous							
DMF	Azeo			16.5	1x/3/1367		
DMAc							
DMSO							
Sulfolane							
CS_2				1.3	1x/1/393		
Acetic acid	83	95	3204				
Aniline	None		11197	6.2	5/189		
Nitrobenzene				2.5	6b/241		
Morpholine							
Pyridine	75	96	8860	2.9	6b/239		
2-Nitropropane	80	95	6289				
Acetonitrile	34	77	2810	31.3	1x/1/393		
Furfuraldehyde	5	98	8781	8.5	3a/137		
Phenol	None		10936	8.9	2b/382		
Water	75	90	734				

n-Nonane

Alternative names

Reference codes

CAS number		Hazchem code	
UN number	1920	EPA code	

Physical properties

Molecular weight	128	Cubic expansion coeff (per °C × 10^3)	1.1
Empirical formula	C_9H_{20}	Surface tension (@20°C dyn/cm)	22.9
Boiling point (°C)	151	Absolute viscosity (@25°C cP)	0.67
Freezing point (°C)	–53	Refractive index (25°C)	1.403
Specific gravity (20/4)	0.718		

Fire hazards

Flash point (closed cup °C)	31	Lower explosive limit (ppm)	8700
Autoignition temperature (°C)	205	Upper explosive limit (ppm)	29000
Electrical conductivity			

Health hazards

IDLH (ppm)		Vapour concentration @21°C ppm	5263
OES-TWA	300	Vapour density (relative to air)	4.4
OES-STEL		Vapour pressure @21°C mmHg	4
Odour threshold (ppm)	0.4	POCP	47

Aqueous effluent

Solubility in water (25°C %w/w)	0.2E–4
Solubility of water in (25°C %w/w)	79E–4
Log_{10} activated carbon partition	
Log_{10} partition in octanol/water (w/w)	
Biological oxygen demand w/w (days)	
Theoretical oxygen demand w/w	3.50

Vapour pressure equation constants (Log_{10}, mmHg)

Antoine equation	A	6.93442
	B	1429.459
	C	201.820
Cox chart	A	7.24534
	B	1662.9

Solvent properties

Solubility parameter		Kauri butanol value	
Dipole (D)		Evaporation time (ether = 1)	
Dielectric constant (20°C)		Evaporation time (BuAc = 1)	
Polarity (water 100)	0.9		

Thermal information

Latent heat (cal/mol)	9037
Nett heat of combustion (kcal/gmol)	1359
Specific heat (cal/mol/°C)	71
Critical pressure (MN/m^2)	2.31
Critical temperature (K)	595
Latent heat of fusion (cal/mol)	3690
Van der Waals' volume	6.52
Van der Waals' surface area	5.48
Molar volume	178.3

Solute	Azeotrope X% w/w	°C	Reference	Solute γ^{∞}	Reference	Partition coefficient	Reference
Hydrocarbons							
n-Pentane				1.0	1x/3/1387		
n-Hexane				0.8	1x/3/1387		
n-Heptane				0.8	1x/3/1387		
n-Octane							
n-Nonane	–		–	–	–	–	–
n-Decane							
2,2,4-TMP							
Cyclohexane							
Benzene				1.0	1x/3/1387		
Toluene							
Ethylbenzene	None		14106				
Xylenes	19	144	14118				
C_9 Aromatics							
Tetralin							
Alcohols							
Methanol	17	64	2119				
Ethanol				6.7	2e/398	6.16	V2/391
n-Propanol	2	97		4.2	2e/505	1.25	V2/586
i-Propanol				9.5	2f/95	2.11	V2/623
n-Butanol	29	116	8203	12.0	2f/209		
i-Butanol							
s-Butanol							
n-Amyl alc.				5.6	1x/3/1387		
i-Amyl alc.							
Cyclohexanol							
1-Octanol							
Ethanediol							
DEG							
1,2-Propanediol							
Glycol ethers							
PGME							
EGME							
EEE	50	128	8493				
EGBE							
Chlorinated							
MDC							
Chloroform							
Carbon tet.							
1,2-EDC							
1,1,1-TCA							
TCE							
Perk.							
MCB							

Hydrocarbons

Solute	Azeotrope X% w/w	Azeotrope °C	Azeotrope Reference	Solute γ^{∞}	Reference	Partition coefficient	Reference
Ketones							
Acetone				6.4	3b/236		
MEK							
MIBK							
Cyclohexanone							
NMP							
Acetophenone				13.7	1x/3/1387		
Ethers							
Diethyl ether							
DIPE							
Dibutyl ether							
MTBE							
1,4-Dioxane				2.2	3+4/481		
THF							
Esters							
Me acetate							
Et acetate							
i-Propyl acetate							
n-Butyl acetate							
Cellosolve acetate							
Miscellaneous							
DMF				16.1	1x/3/1387		
DMAc							
DMSO							
Sulfolane							
CS_2							
Acetic acid	31	113	3230				
Aniline	87	149	11210				
Nitrobenzene							
Morpholine							
Pyridine	10	115	8870	2.4	6b/352		
2-Nitropropane	25	118	6293				
Acetonitrile		80	2812				
Furfuraldehyde							
Phenol							
Water	40	95	789				

n-Decane

Alternative names

Reference codes

CAS number		Hazchem code	
UN number		EPA code	

Physical properties

Molecular weight	142	Cubic expansion coeff (per °C × 10^3)	1.06
Empirical formula	$C_{10}H_{22}$	Surface tension (@20°C dyn/cm)	
Boiling point (°C)	174	Absolute viscosity (@25°C cP)	
Freezing point (°C)	–30	Refractive index (25°C)	1.408
Specific gravity (20/4)	0.730		

Fire hazards

Flash point (closed cup °C)	44	Lower explosive limit (ppm)	8000
Autoignition temperature (°C)		Upper explosive limit (ppm)	25000
Electrical conductivity			

Health hazards

IDLH (ppm)		Vapour concentration @21°C ppm	1980
OES-TWA		Vapour density (relative to air)	4.9
OES-STEL		Vapour pressure @21°C mmHg	1.5
Odour threshold (ppm)		POCP	46.4

Aqueous effluent

Solubility in water (25°C %w/w)	0.02E–4
Solubility of water in (25°C %w/w)	72E–4
Log_{10} activated carbon partition	
Log_{10} partition in octanol/water (w/w)	
Biological oxygen demand w/w (days)	
Theoretical oxygen demand w/w	3.49

Vapour pressure equation constants (Log_{10}, mmHg)

Antoine equation	A	7.44000
	B	1843.12
	C	230.220
Cox chart	A	7.33946
	B	1801.3

Solvent properties

Solubility parameter	6.7	Kauri butanol value	
Dipole (D)		Evaporation time (ether = 1)	
Dielectric Constant (20°C)		Evaporation time (BuAc = 1)	
Polarity (water 100)	0.9		

Thermal information

Latent heat (cal/mol)	
Nett heat of combustion (kcal/gmol)	1504
Specific heat (cal/mol/°C)	
Critical pressure (MN/m^2)	2.11
Critical temperature (K)	617
Latent heat of fusion (cal/mol)	6864
Van der Waals' volume	7.20
Van der Waals' surface area	6.02
Molar volume	195

Solute	Azeotrope X% w/w	°C	Reference	Solute γ^{∞}	Reference	Partition coefficient	Reference
Hydrocarbons							
n-Pentane				1.0	1x/4/1410		
n-Hexane				1.0	1x/4/1411		
n-Heptane				1.0	1x/4/1412		
n-Octane				1.0	1x/4/1412		
n-Nonane							
n-Decane	–		–	–	–	–	–
2,2,4-TMP							
Cyclohexane				0.9	1x/4/1411		
Benzene				0.8	6c/574		
Toluene				1.3	1x/4/1411		
Ethylbenzene							
Xylenes							
C_9 Aromatics	75	166	15033				
Tetralin							
Alcohols							
Methanol	None		2126	0.6	2e/193		
Ethanol	None			3.3	2a/508		
n-Propanol	None			5.8	2a/606		
i-Propanol	None			5.1	2b/118		
n-Butanol	92		8213	12.4	2b/236		
i-Butanol							
s-Butanol	None			1.1	2d/285	0.06	V3/133
n-Amyl alc.							
i-Amyl alc.							
Cyclohexanol							
1-Octanol				7.1	2f/542		
Ethanediol	77	161	4434				
DEG							
1,2-Propanediol							
Glycol ethers							
PGME							
EGME	8	123	6628				
EEE							
EGBE							
Chlorinated							
MDC							
Chloroform							
Carbon tet.							
1,2-EDC							
1,1,1-TCA							
TCE							
Perk.							
MCB				1.3	6b/392		

Hydrocarbons

Solute	Azeotrope X% w/w	Azeotrope °C	Azeotrope Reference	Solute γ^∞	Reference	Partition coefficient	Reference
Ketones							
Acetone	None		5396	1.2	3+4/247		
MEK	None			3.8	3b/396		
MIBK							
Cyclohexanone							
NMP				12.6	1x/4/1410		
Acetophenone							
Ethers							
Diethyl ether							
DIPE							
Dibutyl ether							
MTBE							
1,4-Dioxane				3.2	1x/4/1410		
THF							
Esters							
Me acetate							
Et acetate							
i-Propyl acetate							
n-Butyl acetate							
Cellosolve acetate							
Miscellaneous							
DMF				15.5	1x/4/1410		
DMAc							
DMSO							
Sulfolane							
CS_2							
Acetic acid	20	117	3237	1.1	6c/571		
Aniline	64	167	11230	5.3	5/191		
Nitrobenzene							
Morpholine							
Pyridine	None		8872	1.7	6b/386		
2-Nitropropane							
Acetonitrile		81	2815				
Furfuraldehyde				4.5	3+4/59		
Phenol	65	168	11016				
Water		97	842				

2,2,4-Trimethyl pentane

Alternative names
ISO-Octane, ISOPAR C, 2,2,4-TMP

Reference codes

CAS number	540 84 1	Hazchem code	3YE
UN number		EPA code	

Physical properties

Molecular weight	114	Cubic expansion coeff (per °C × 10^3)	1.00
Empirical formula	C_8H_{18}	Surface tension (@20°C dyn/cm)	18.33
Boiling point (°C)	99	Absolute viscosity (@25°C cP)	0.477
Freezing point (°C)	–107	Refractive index (25°C)	1.389
Specific gravity (20/4)	0.692		

Fire hazards

Flash point (closed cup °C)	–12	Lower explosive limit (ppm)	11000
Autoignition temperature (°C)	418	Upper explosive limit (ppm)	60000
Electrical conductivity			

Health hazards

IDLH (ppm)		Vapour concentration @21°C ppm	
OES-TWA	400	Vapour density (relative to air)	4.1
OES-STEL		Vapour pressure @21°C mmHg	41
Odour threshold (ppm)		POCP	

Aqueous effluent

Solubility in water (25°C %w/w)	2.2E–4
Solubility of water in (25°C %w/w)	0.011
Log_{10} activated carbon partition	
Log_{10} partition in octanol/water (w/w)	
Biological oxygen demand w/w (days)	
Theoretical oxygen demand w/w	3.51

Vapour pressure equation constants (Log_{10}, mmHg)

Antoine equation	A	6.80304
	B	1252.59
	C	220.119
Cox chart	A	7.04642
	B	1370.5

Solvent properties

Solubility parameter	7.4	Kauri butanol value	27
Dipole (D)		Evaporation time (ether = 1)	
Dielectric constant (20°C)		Evaporation time (BuAc = 1)	
Polarity (water 100)			

Thermal information

Latent heat (cal/mol)	7396
Nett heat of combustion (kcal/gmol)	1211
Specific heat (cal/mol/°C)	55.6
Critical pressure (MN/m^2)	2.59
Critical temperature (K)	544
Latent heat of fusion (cal/mol)	2157
Van der Waals' volume	5.85
Van der Waals' surface area	5.01
Molar volume	166.1

Solute	Azeotrope X% w/w	°C	Reference	Solute γ^{∞}	Reference	Partition coefficient	Reference
Hydrocarbons							
n-Pentane							
n-Hexane				1.0	1x/1/398		
n-Heptane	None		13809	1.0	6b/197		
n-Octane	None		14715	0.7	6b/283		
n-Nonane							
n-Decane							
2,2,4-TMP	–		–	–	–	–	–
Cyclohexane	None		11699a	1.9	6a/328		
Benzene	2	80	10880	1.4	6b/304		
Toluene	None		13040	1.4	6b/323		
Ethylbenzene	None			1.4	6b/333		
Xylenes							
C_9 Aromatics							
Tetralin							
Alcohols							
Methanol	47	59	2114	16.6	2a/250	20.1	V2/139
Ethanol	58	72	4167	29.3	2a/503	31.2	V2/386
n-Propanol	59	85	6524	11.2	2c/500		
i-Propanol	46	77	6419	4.4	2b/116		
n-Butanol							
i-Butanol	73	92	8378				
s-Butanol	66	88	8254	5.2	2b/284		
n-Amyl alc.							
i-Amyl alc.	95	99	9882				
Cyclohexanol							
1-Octanol							
Ethanediol							
DEG							
1,2-Propanediol							
Glycol ethers							
PGME							
EGME							
EEE							
EGBE							
Chlorinated							
MDC				2.1	1x/1/397		
Chloroform				1.5	1x/1/397		
Carbon tet.	None		1168b	1.2	6b/285		
1,2-EDC							
1,1,1-TCA							
TCE							
Perk.							
MCB							

Hydrocarbons

Solute	Azeotrope X% w/w	Azeotrope °C	Azeotrope Reference	Solute γ^{∞}	Reference	Partition coefficient	Reference
Ketones							
Acetone				2.4	3b/225	0.11	V3/30
MEK	30			3.7	3b/395	0.10	CEH
MIBK							
Cyclohexanone							
NMP							
Acetophenone							
Ethers							
Diethyl ether							
DIPE							
Dibutyl ether							
MTBE							
1,4-Dioxane							
THF				1.4	1x/1/398		
Esters							
Me acetate							
Et acetate				3.1	1x/1/398		
i-Propyl acetate							
n-Butyl acetate							
Cellosolve acetate							
Miscellaneous							
DMF							
DMAc							
DMSO							
Sulfolane							
CS_2				1.3	1x/1/397		
Acetic acid							
Aniline	None		11196	6.5	6b/318		
Nitrobenzene							
Morpholine							
Pyridine	77	96	8868	7.5	6b/297		
2-Nitropropane	79	95	6292				
Acetonitrile	40	69	2811	31.5	1x/1/397		
Furfuraldehyde	Azeo			13.1	3+4/55		
Phenol	None			17.3	2b/383		
Water	89	79	734a				

Cyclohexane

Alternative names
Hexamethylene, benzene hydride

Reference codes

CAS number	110 82 7	Hazchem code	3YE
UN number	1145	EPA code	

Physical properties

Molecular weight	84	Cubic expansion coeff (per °C × 10^3)	1.2
Empirical formula	C_6H_{12}	Surface tension (@20°C dyn/cm)	24.98
Boiling point (°C)	81	Absolute viscosity (@25°C cP)	0.980
Freezing point (°C)	+6.5	Refractive index (25°C)	1.424
Specific gravity (20/4)	0.778		

Fire hazards

Flash point (closed cup °C)	–17	Lower explosive limit (ppm)	13000
Autoignition temperature (°C)	260	Upper explosive limit (ppm)	84000
Electrical conductivity	7.0E–18		

Health hazards

IDLH (ppm)	10000	Vapour concentration @21°C ppm	155700
OES-TWA	100	Vapour density (relative to air)	2.9
OES-STEL	300	Vapour pressure @21°C mmHg	78.8
Odour threshold (ppm)	400	POCP	25

Aqueous effluent

Solubility in water (25°C %w/w)	0.0055
Solubility of water in (25°C %w/w)	0.01
Log_{10} activated carbon partition	
Log_{10} partition in octanol/water (w/w)	+4.15
Biological oxygen demand w/w (days)	0.6 (5)
Theoretical oxygen demand w/w	3.43

Vapour pressure equation constants (Log_{10}, mmHg)

Antoine equation	A	6.85146
	B	1206.470
	C	223.136
Cox chart	A	7.04736
	B	1295.8

Solvent properties

Solubility parameter	8.2	Kauri butanol value	50
Dipole (D)	0.3	Evaporation time (ether = 1)	3.4
Dielectric Constant (20°C)	2.01	Evaporation time (BuAc = 1)	5.6
Polarity (water 100)	0.6		

Thermal information

Latent heat (cal/mol)	7140
Nett heat of combustion (kcal/gmol)	874
Specific heat (cal/mol/°C)	36.4
Critical pressure (MN/m^2)	4.07
Critical temperature (K)	553
Latent heat of fusion (cal/mol)	627
Van der Waals' volume	4.05
Van der Waals' surface area	3.24
Molar volume	108.57

Solute	Azeotrope X% w/w	Azeotrope °C	Reference	Solute γ^{∞}	Reference	Partition coefficient	Reference
Hydrocarbons							
n-Pentane				1.2	6a/119		
n-Hexane	None		11690	4.0	6a/273		
n-Heptane	None		11697	0.9	6a/304		
n-Octane				1.7	6a/323		
n-Nonane							
n-Decane							
2,2,4-TMP	None		11699a	1.0	6a/328		
Cyclohexane	–		–	–	–	–	–
Benzene	50	77	10854	1.4	6a/205	<0.01	V3/278
Toluene	None		11694	1.5	6a/283	<0.01	V3/340
Ethylbenzene				1.7	6a/310		
Xylenes				2.0	6a/311		
C_9 Aromatics							
Tetralin							
Alcohols							
Methanol	62	54	2079	20.9	2a/239	14.8	P159
Ethanol	70	65	4087	13.0	2a/430	23.0	V2/350
n-Propanol	80	74	6495	10.0	2a/579	6.6	P639
i-Propanol	68	69	6384	12.0	2f/69	8.0	V2/613
n-Butanol	96	80	8146	9.0	2f/179	1.8	P950
i-Butanol	86	78	8338	9.9	2f/317		
s-Butanol	82	76	8234	4.5	2f/234		
n-Amyl alc.	None		9752	16.7	1x/3/1227		
i-Amyl alc.	None		9834				
Cyclohexanol	None		11684				
1-Octanol							
Ethanediol	None		4255				
DEG							
1,2-Propanediol							
Glycol ethers							
PGME							
EGME	85	77	6572	22.1	2b/128		
EEE	None		8430				
EGBE						1.4	P3977
Chlorinated							
MDC							
Chloroform	None		1490				
Carbon tet.	None		1159	1.1	6a/142		
1,2-EDC	50	74	3001	2.6	6a/159		
1,1,1-TCA							
TCE	None		2328	1.5	6a/155		
Perk.							
MCB	None		10515	1.71	6a/202		

Hydrocarbons

Solute	Azeotrope X% w/w	°C	Reference	Solute γ^∞	Reference	Partition coefficient	Reference
Ketones							
Acetone	33	53	5378	4.7	3+4/213	1.95	P492
MEK	60	72	7374	2.8	3+4/297	0.13	V3/25
MIBK	None		11685	2.7	3+4/354		
Cyclohexanone				2.7	3+4/337		
NMP				9.4	3b/447		
Acetophenone							
Ethers							
Diethyl ether							
DIPE				1.0	3+4/555		
Dibutyl ether							
MTBE				1.3	1x/3/1227		
1,4-Dioxane	75	80	7540	3.0	3+4/468		
THF				1.6	1x/3/1227		
Esters							
Me acetate	17	55	5341	3.4	5/393		
Et acetate	46	73	7583	2.7	5/506		
i-Propyl acetate	75	79	9296				
n-Butyl acetate				2.1	5/585		
Cellosolve acetate							
Miscellaneous							
DMF				19.2	1x/3/1226		
DMAc				11.2	1x/1/271		
DMSO							
Sulfolane				90.0	1x/3/1227		
CS_2	None		1269	1.1	6a/154		
Acetic acid	9	79	3173	15.6	5/146	12.8	V2/259
Aniline	None		11148	4.7	6a/255	0.2	P1713
Nitrobenzene				9.9	6a/203		
Morpholine							
Pyridine	None		8846	2.8	6a/177		
2-Nitropropane	90	81	6283	6.9	1x/1/270		
Acetonitrile	40	62	2797				
Furfuraldehyde	None		8763	16.7	3+4/45		
Phenol						1.59	P1622
Water	91	70	522				

Benzene

Alternative names
Benzole, benzol, **not** benzine

Reference codes

CAS number	71 43 2	Hazchem code	3WE
UN number	1114	EPA code	U019

Physical properties

Molecular weight	78	Cubic expansion coeff (per °C × 10^3)	1.24
Empirical formula	C_6H_6	Surface tension (@20°C dyn/cm)	28.9
Boiling point (°C)	80	Absolute viscosity (@25°C cP)	0.65
Freezing point (°C)	+5.5	Refractive index (25°C)	1.498
Specific gravity (20/4)	0.879		

Fire hazards

Flash point (closed cup °C)	–11	Lower explosive limit (ppm)	13000
Autoignition temperature (°C)	592	Upper explosive limit (ppm)	71000
Electrical conductivity	4.4E–17		

Health hazards

IDLH (ppm)	2000	Vapour concentration @21°C ppm	117000
OES-TWA	5	Vapour density (relative to air)	2.8
OES-STEL		Vapour pressure @21°C mmHg	78
Odour threshold (ppm)	300	POCP	19

Aqueous effluent

Solubility in water (25°C %w/w)	0.18
Solubility of water in (25°C %w/w)	0.063
Log_{10} activated carbon partition	3.6
Log_{10} partition in octanol/water (w/w)	+2.13
Biological oxygen demand w/w (days)	1.2 (10)
Theoretical oxygen demand w/w	3.08

Vapour pressure equation constants (Log_{10}, mmHg)

Antoine equation	A	6.87987
	B	1196.760
	C	219.161
Cox chart	A	7.04500
	B	1290.9

Solvent properties

Solubility parameter	9.2	Kauri butanol value	112
Dipole (D)	0	Evaporation time (ether = 1)	2.6
Dielectric constant (20°C)	2.28	Evaporation time (BuAc = 1)	
Polarity (water 100)	11.1		

Thermal information

Latent heat (cal/mol)	7340
Nett heat of combustion (kcal/gmol)	749
Specific heat (cal/mol/°C)	31
Critical pressure (MN/m^2)	4.9
Critical temperature (K)	562
Latent heat of fusion (cal/mol)	2375
Van der Waals' volume	3.19
Van der Waals' surface area	2.40
Molar volume	89.41

Solute	Azeotrope X% w/w	°C	Reference	Solute γ^{∞}	Reference	Partition coefficient	Reference
Hydrocarbons							
n-Pentane	None		9741	1.8	6a/118		
n-Hexane	5	68	10861	1.5	6a/535		
n-Heptane	0.7	80	10876	1.7	6b/123		
n-Octane	None		10879	1.2	6b/242		
n-Nonane							
n-Decane	None			1.1	6c/574		
2,2,4-TMP	98	80	10880	1.5	6b/304		
Cyclohexane	50	77	10854	1.5	6a/205		
Benzene	–		–	–	–	–	–
Toluene	None		10871	0.9	7/823		
Ethylbenzene	None		10878a	1.0	7/306		
Xylenes	None			1.0	7/310		
C_9 Aromatics	None		10882a	2.2	7/322		
Tetralin							
Alcohols							
Methanol	61	57	2066	9.7	2a/205	2.3	V2/121
Ethanol	68	68	4073	13.2	2a/399	8.0	P379
n-Propanol	83	77	6491	5.7	2a/556	1.3	P636
i-Propanol	67	72	6375	6.9	2f/65	0.86	V2/595
n-Butanol	None		8136	4.3	2f/169	0.19	V3/118
i-Butanol	92	79	8333	116.0	2f/316	0.24	V3/142
s-Butanol	85	79	8232	3.6	2f/227	0.33	V3/129
n-Amyl alc.	None		9748			0.15	P1268
i-Amyl alc.	None		9821				
Cyclohexanol	None		10856				
1-Octanol							
Ethanediol	None		4239				
DEG	None		8520				
1,2-Propanediol	None		6654				
Glycol ethers							
PGME							
EGME	None		6567	5.7	2b/127		
EEE	None		8425				
EGBE						1.42	P3978
Chlorinated							
MDC				0.9	1x/1/225		
Chloroform	None		1486	0.9	7/80		
Carbon tet.	None		1154	1.1	7/7		
1,2-EDC	82	80	2999	1.0	7/142		
1,1,1-TCA				1.0	7/121		
TCE	None		2326	1.0	7/114		
Perk.	None			1.3	7/112		
MCB	None		10509	1.0	7/243		

Hydrocarbons

Solute	Azeotrope X% w/w	Azeotrope °C	Azeotrope Reference	Solute γ^∞	Reference	Partition coefficient	Reference
Ketones							
Acetone	None		5374	1.9	3+4/195	0.43	P497
MEK	55	78	7369	1.3	3+4/284		
MIBK	None		10857	1.1	3+4/351		
Cyclohexanone				0.7	3b/503		
NMP				1.0	3b/441		
Acetophenone						0.001	P2745
Ethers							
Diethyl ether	None		8293	0.7	3+4/516		
DIPE	None		10863	1.2	3+4/553		
Dibutyl ether							
MTBE							
1,4-Dioxane	88	12	7537	1.1	3+4/465	0.24	V3/72
THF				0.8	1x/3/1183	0.06	V4/238
Esters							
Me acetate	0.3	43	5537	1.3	5/375	0.08	P518
Et acetate	6	77	7580	3.7	5/502	0.02	P863
i-Propyl acetate	None		9294				
n-Butyl acetate	None		10859	0.9	5/583	<0.01	V3/279
Cellosolve acetate							
Miscellaneous							
DMF	None		5863	4.6	1x/1/226	3.33	V2/542
DMAc				3.4	1x/1/227		
DMSO	None		4184	3.3	7/169		
Sulfolane				3.8	7/191	0.53	V3/88
CS_2	None		1265	1.3	7/100		
Acetic acid	98	80	3163	5.1	5/127	21.0	P301
Aniline	None		10850	1.5	7/263	0.12	P1716
Nitrobenzene	None		10703	1.5	7/253		
Morpholine						2.49	V3/105
Pyridine	None		8841	1.2	7/220	0.09	P1104
2-Nitropropane	None		6281	2.2	7/186		
Acetonitrile	66	73	2795	3.0	7/124	0.03	V2/182
Furfuraldehyde	None		8760	1.8	3+4/44	0.05	V3/190
Phenol	None			4.8	2b/359	0.08	V3/265
Water	91	69	486				

Toluene

Alternative names
Toluol, methylbenzene, methylbenzol

Reference codes

CAS number	108 88 3	Hazchem code	3YE
UN number	1294	EPA code	U220

Physical properties

Molecular weight	92	Cubic expansion coeff (per °C × 10^3)	1.1
Empirical formula	C_7H_8	Surface tension (@20°C dyn/cm)	28.5
Boiling point (°C)	110.6	Absolute viscosity (@25°C cP)	0.59
Freezing point (°C)	–95	Refractive index (25°C)	1.494
Specific gravity (20/4)	0.867		

Fire hazards

Flash point (closed cup °C)	4	Lower explosive limit (ppm)	12700
Autoignition temperature (°C)	480	Upper explosive limit (ppm)	70000
Electrical conductivity	8.0E–16		

Health hazards

IDLH (ppm)	2000	Vapour concentration @21°C ppm	31000
OES-TWA	50	Vapour density (relative to air)	3.2
OES-STEL	150	Vapour Pressure @21°C mmHg	23.2
Odour threshold (ppm)	40	POCP	56

Aqueous effluent

Solubility in water (25°C %w/w)	0.052
Solubility of water in (25°C %w/w)	0.033
Log_{10} activated carbon partition	2.9
Log_{10} partition in octanol/water (w/w)	+2.69
Biological oxygen demand w/w (days)	1.19 (5)
Theoretical oxygen demand w/w	3.13

Vapour pressure equation constants (Log_{10}, mmHg)

Antoine equation	A	6.95087
	B	1342.31
	C	219.187
Cox chart	A	7.12773
	B	1448.2

Solvent properties

Solubility parameter	8.9	Kauri butanol value	105
Dipole (D)	0.4	Evaporation time (ether = 1)	6.1
Dielectric Constant (20°C)	2.38	Evaporation time (BuAc = 1)	2.0
Polarity (water 100)	9.9		

Thermal information

Latent heat (cal/mol)	7985
Nett heat of combustion (kcal/gmol)	892
Specific heat (cal/mol/°C)	41.0
Critical pressure (MN/m^2)	4.22
Critical temperature (K)	591.8
Latent heat of fusion (cal/mol)	1580
Van der Waals' volume	3.92
Van der Waals' surface area	2.97
Molar volume	106.85

Solute	Azeotrope X% w/w	°C	Reference	Solute γ^{∞}	Reference	Partition coefficient	Reference
Hydrocarbons							
n-Pentane				1.3	6c/160		
n-Hexane	None		12131	1.4	6a/591		
n-Heptane	None		13027	1.2	6b/169		
n-Octane	None		13041	1.3	6b/261		
n-Nonane							
n-Decane							
2,2,4-TMP	None		13043	1.4	6b/323		
Cyclohexane	None		11694	1.2	6a/283		
Benzene	None		10871	1.0	7/283		
Toluene	–		–	–	–	–	–
Ethylbenzene	None		13029	1.1	7/443		
Xylenes	None		13030	0.83	7/444		
C_9 Aromatics							
Tetralin							
Alcohols							
Methanol	31	64	2098	6.3	2a/268	20.8	V2/135
Ethanol	32	77	4120	5.3	2a/477	11.6	V2/372
n-Propanol	51	93	6512	4.3	2a/592	1.05	V2/580
i-Propanol	31	81	6397	3.9	2f/78	1.16	V2/619
n-Butanol	68	106	8170	10.3	2f/190	0.17	CEH
i-Butanol	55	101	8361	3.8	2b/289		
s-Butanol	45	95	8246	3.3	2d/276	0.20	V3/130
n-Amyl alc.	None		9760				
i-Amyl alc.	90	110	9852				
Cyclohexanol	None		11720				
1-Octanol							
Ethanediol	93	110	4285				
DEG	None		8531	45.9	2f/341		
1,2-Propanediol	98	110	6658				
Glycol ethers							
PGME	None		9978				
EGME	74	106	6586				
EEE	89	110	8450	3.9	2f/337		
EGBE	None			1.4	2f/440	0.78	P3980
Chlorinated							
MDC				0.9	1x/3/317		
Chloroform	None		1498	1.2	7/352		
Carbon tet.	None		1166	1.0	7/332		
1,2-EDC	None		3006	1.0	7/380		
1,1,1-TCA							
TCE				0.9	7/370		
Perk.	None		2220				
MCB	None		10524	1.0	7/416		

Hydrocarbons

Solute	Azeotrope X% w/w	°C	Reference	Solute γ^∞	Reference	Partition coefficient	Reference
Ketones							
Acetone	None		5391	1.8	3+4/236	0.40	P501
MEK	None		7382	1.4	3+4/308		
MIBK	97	111	11799	1.2	3+4/356		
Cyclohexanone				1.4	3+4/339		
NMP				0.3	3b/456		
Acetophenone							
Ethers							
Diethyl ether	None		8304				
DIPE				1.2	3+4/558		
Dibutyl ether							
MTBE							
1,4-Dioxane	20	102	7550	1.3	3+4/375		
THF							
Esters							
Me acetate							
Et acetate	None		7591	1.2	5/516		
i-Propyl acetate	None		9300				
n-Butyl acetate	None		11825	1.1	5/586		
Cellosolve acetate	None		11913				
Miscellaneous							
DMF	None		5893a	1.9	7/390	2.22	V4/223
DMAc							
DMSO	None			8.3	7/386		
Sulfolane				140.4	7/399		
CS_2	None		1276	1.1	7/361		
Acetic acid	72	101	3194	9.4	5/159	2.97	V4/190
Aniline	None		11167	1.7	7/426	0.03	P1719
Nitrobenzene	None		10718	1.9	7/422		
Morpholine							
Pyridine	68	108	8858	1.8	7/406	0.14	P1109
2-Nitropropane	82	110	6285				
Acetonitrile	24	81	2801	3.8	7/373	0.07	V2/191
Furfuraldehyde	None		8776	2.6	3a/135	0.03	V4/258
Phenol	None		10920	2.4	2f/393	0.12	P1644
Water	80	85		610			

Ethylbenzene

Alternative names
Phenyl ethane

Reference codes

CAS number	100 41 4	Hazchem code	3YE
UN number	1175	EPA code	Z048

Physical properties

Molecular weight	106	Cubic expansion coeff (per °C × 10^3)	1.03
Empirical formula	C_8H_{10}	Surface tension (@20°C dyn/cm)	29.2
Boiling point (°C)	136	Absolute viscosity (@25°C cP)	0.72
Freezing point (°C)	–94	Refractive index (25°C)	1.493
Specific gravity (20/4)	0.867		

Fire hazards

Flash point (closed cup °C)	15	Lower explosive limit (ppm)	10000
Autoignition temperature (°C)	435	Upper explosive limit (ppm)	67000
Electrical conductivity			

Health hazards

IDLH (ppm)	2000	Vapour concentration @21°C ppm	9960
OES-TWA	100	Vapour density (relative to air)	3.7
OES-STEL	125	Vapour pressure @21°C mmHg	8.0
Odour threshold (ppm)	125	POCP	59

Aqueous effluent

Solubility in water (25°C %w/w)	0.020
Solubility of water in (25°C %w/w)	0.033
Log_{10} activated carbon partition	3.1
Log_{10} partition in octanol/water (w/w)	+2.76
Biological oxygen demand w/w (days)	0.028 (5)
Theoretical oxygen demand w/w	3.17

Vapour pressure equation constants (Log_{10}, mmHg)

Antoine equation	A	6.96580
	B	1429.550
	C	213.767
Cox chart	A	7.19691
	B	1579.7

Solvent properties

Solubility Parameter	8.9	Kauri butanol value	
Dipole (D)	0.4	Evaporation time (ether = 1)	8.8
Dielectric constant (20°C)	2.41	Evaporation time (BuAc = 1)	0.84
Polarity (water 100)			

Thermal information

Latent heat (cal/mol)	8491
Nett heat of combustion (kcal/gmol)	1038
Specific heat (cal/mol/°C)	43
Critical pressure (MN/m^2)	3.74
Critical temperature (K)	617
Latent heat of fusion (cal/mol)	
Van der Waals' volume	4.60
Van der Waals' surface area	3.51
Molar volume	123.1

Solute	Azeotrope X% w/w	Azeotrope °C	Reference	Solute γ^{∞}	Reference	Partition coefficient	Reference
Hydrocarbons							
n-Pentane							
n-Hexane							
n-Heptane	None		13807	1.7	6c/489		
n-Octane	12	126	14101	1.2	6b/273		
n-Nonane	None		14106				
n-Decane							
2,2,4-TMP	None			1.5	6b/333		
Cyclohexane				1.2	6a/310		
Benzene	None		10878a	1.0	7/306		
Toluene	None		13029	1.0	7/443		
Ethylbenzene	–		–	–	–	–	–
Xylenes	None		14098				
C_9 Aromatics							
Tetralin							
Alcohols							
Methanol	None		2106	6.5	2c/245		
Ethanol	None		4144	6.1	2c/460		
n-Propanol	9	97	6517	3.3	2a/601		
i-Propanol	None		6402	2.0	2d/95	2.0	V2/621
n-Butanol	33	115	8185	2.7	2b/228		
i-Butanol	20	107	8371				
s-Butanol	None		8251				
n-Amyl alc.	60	130	9765				
i-Amyl alc.	51	126	9871				
Cyclohexanol							
1-Octanol							
Ethanediol	87	133	4321				
DEG	Azeo		8545				
1,2-Propanediol							
Glycol ethers							
PGME	80	135	9991				
EGME	49	117	6596	3.1	2b/132		
EEE	57	126	8463	2.9	2b/299	2.0	V3/158
EGBE	4	140				0.85	P3977
Chlorinated							
MDC							
Chloroform							
Carbon tet.	None		1167a	0.9	7/464		
1,2-EDC				0.7	7/466		
1,1,1-TCA							
TCE							
Perk.	None		2225				
MCB				1.0	7/469		

Solute	Azeotrope X% w/w	Azeotrope °C	Azeotrope Reference	Solute γ^{∞}	Reference	Partition coefficient	Reference
Ketones							
Acetone	None			2.1	3b/217		
MEK	None			1.4	3+4/316		
MIBK	None		11802				
Cyclohexanone							
NMP							
Acetophenone							
Ethers							
Diethyl ether							
DIPE				1.1	3+4/563		
Dibutyl ether	None		14102				
MTBE							
1,4-Dioxane							
THF							
Esters							
Me acetate							
Et acetate				0.9	5/540		
i-Propyl acetate							
n-Butyl acetate	None		11828				
Cellosolve acetate	None		12007				
Miscellaneous							
DMF	85	134					
DMAc							
DMSO							
Sulfolane							
CS_2							
Acetic acid	34	115	3206	5.0	5/178		
Aniline	None		11183	2.0	7/474		
Nitrobenzene				1.6	7/470		
Morpholine				1.5	3+4/482		
Pyridine	None		8861				
2-Nitropropane	8	120	6290				
Acetonitrile	None		2804	2.5	7/465		
Furfuraldehyde	95	132	8783	3.0	3+4/51		
Phenol	None		10943			0.07	V4/268
Water	67	92	676				

Xylene (mixed isomers)

Alternative names
Dimethyl benzenes, xylol

Reference codes

CAS number	1330 20 7	Hazchem code	3Y
UN number	1307	EPA code	U239

*Physical properties**

Molecular weight	106	Cubic expansion coeff (per °C x 10^3)	1.0
Empirical formula	C_8H_{10}	Surface tension (@20°C dyn/cm)	28.6
Boiling point (°C)	136*	Absolute viscosity (@25°C cP)	0.7*
Freezing point (°C)		Refractive index (25°C)	1.496
Specific gravity (20/4)	0.870		

Fire hazards

Flash point (closed cup °C)	23*	Lower explosive limit (ppm)	11400
Autoignition temperature (°C)	480	Upper explosive limit (ppm)	70000
Electrical conductivity	8.0E–16		

Health hazards

IDLH (ppm)	10000	Vapour concentration @21°C ppm	9180
OES-TWA	100	Vapour density (relative to air)	3.7
OES-STEL	150	Vapour pressure @ 21°C mmHg	7.0
Odour threshold (ppm)	1.0	POCP	85*

Aqueous effluent

Solubility in water (25°C %w/w)	0.02
Solubility of water in (25°C %w/w)	0.05
Log_{10} activated carbon partition	4.3
Log_{10} partition in octanol/water (w/w)	3.0
Biological oxygen demand w/w (days)	0.1 (5)
Theoretical oxygen demand w/w	3.17

*Vapour pressure equation constants (Log_{10}, mmHg)**

Antoine equation	A	6.99053
	B	1453.43
	C	215.310
Cox chart	A	7.20807
	B	1601.1

Solvent properties

Solubility parameter	8.9	Kauri butanol value	98
Dipole (D)	1.3	Evaporation time (ether = 1)	13.5
Dielectric constant (20°C)	2.3	Evaporation time (BuAc = 1)	0.76
Polarity (water 100)	7.4		

Thermal information

Latent heat (cal/mol)	8692
Nett heat of combustion (kcal/gmol)	1035
Specific heat (cal/mol/°C)	42
Critical pressure (MN/m^2)	3.55
Critical temperature (K)	623
Latent heat of fusion (cal/mol)	3180
Van der Waals' volume	4.66
Van der Waals' surface area	3.54
Molar volume	121.84

*Typical mixture of isomers

Solute	Azeotrope X% w/w	Azeotrope °C	Reference	Solute γ^∞	Reference	Partition coefficient	Reference
Hydrocarbons							
n-Pentane				1.5	1x/1/369		
n-Hexane	None			1.2	6a/605		
n-Heptane	None		13808	1.4	6c/497		
n-Octane	None		14120	1.3	6b/275		
n-Nonane	81	144	14118				
n-Decane							
2,2,4-TMP							
Cyclohexane				1.5	6a/311		
Benzene	None			1.0	7/310		
Toluene	None		13030	1.1	7/444		
Ethylbenzene	None		14098				
Xylenes	–		–	–	–	–	–
C_9 Aromatics	None						
Tetralin							
Alcohols							
Methanol	None		2108	6.6	2c/247		
Ethanol	None		4146	5.3	2a/500	3.08	V2/379
n-Propanol	7	97	6519	3.3	2c/575		
i-Propanol	None		6402	2.6	2d/229		
n-Butanol	27	115	8186	3.1	2b/229		
i-Butanol	12	108		2.6	2b/292		
s-Butanol	None		8252	2.7	2d/282		
n-Amyl alc.							
i-Amyl alc.							
Cyclohexanol	90	140	11730				
1-Octanol	None	11	4113	6.3	2f/536		
Ethanediol	93	135	4323				
DEG	Azeo						
1,2-Propanediol							
Glycol ethers							
PGME	73	137	9994				
EGME	45	120	6598	4.0	2b/134		
EEE	50	128	8465	4.2	2f/416		
EGBE	4	144	12235			0.56	P3979
Chlorinated							
MDC				0.85	1x/1/369		
Chloroform							
Carbon tet.				0.9	7/480		
1,2-EDC				1.1	7/490		
1,1,1-TCA							
TCE							
Perk.	None		1167c				
MCB				0.9	7/508		

Hydrocarbons

Solute	Azeotrope X% w/w	°C	Reference	Solute γ^∞	Reference	Partition coefficient	Reference
Ketones							
Acetone	None			2.4	3b/222	0.29	V2/506
MEK				1.2	3b/382		
MIBK	None			1.3	3b/553		
Cyclohexanone				1.2	3b/511		
NMP				1.9	3b/462		
Acetophenone							
Ethers							
Diethyl ether				1.1	1x/3/1348		
DIPE							
Dibutyl ether	22	142	14117				
MTBE							
1,4-Dioxane				1.3	1x/3/1350		
THF							
Esters							
Me acetate							
Et acetate	None		7594	1.6	5/541		
i-Propyl acetate							
n-Butyl acetate	None		11829				
Cellosolve acetate	None		12009				
Miscellaneous							
DMF	80	136	5894	2.5	7/481		
DMAc							
DMSO							
Sulfolane							
CS_2				1.1	1x/1/369		
Acetic acid	22	116	3208	3.8	5/181	14.1	P310
Aniline	None		11185			0.11	P1718
Nitrobenzene							
Morpholine							
Pyridine				1.4	7/482	0.08	P1108
2-Nitropropane							
Acetonitrile	None		2805	5.1	1x/1/369	0.38	V2/193
Furfuraldehyde	90	139	8785	2.9	3+4/52		
Phenol	None		10944			0.12	P1642
Water	63	93	677				

C_9 Aromatics

Alternative names

Reference codes

CAS number		Hazchem code	3(Y)
UN number	2325	EPA code	

Physical properties

Molecular weight	120	Cubic expansion coeff (per °C × 10^3)	0.89
Empirical formula	C_9H_{12}	Surface tension (@20°C dyn/cm)	28
Boiling point (°C)	152*	Absolute viscosity (@25°C cP)	1.153
Freezing point (°C)		Refractive index (25°C)	
Specific gravity (20/4)	0.876		

Fire hazards

Flash point (closed cup °C)	55	Lower explosive limit (ppm)	9000
Autoignition temperature (°C)	425	Upper explosive limit (ppm)	65000
Electrical conductivity			

Health hazards

IDLH (ppm)	8000	Vapour concentration @21°C ppm	5200
OES-TWA	25	Vapour density (relative to air)	4.2
OES-STEL	75	Vapour pressure @21°C mmHg	4
Odour threshold (ppm)	0.4	POCP	84*

Aqueous effluent

Solubility in water (25°C %w/w)	0.0058*
Solubility of water in (25°C %w/w)	0.035*
Log_{10} activated carbon partition	
Log_{10} partition in octanol/water (w/w)	
Biological oxygen demand w/w (days)	
Theoretical oxygen demand w/w	3.20

*Vapour pressure equation constants (Log_{10}, mmHg)**

Antoine equation	A	7.0764
	B	1571.0
	C	209.7
Cox chart	A	7.25282
	B	1670.2

Solvent properties

Solubility parameter		Kauri butanol value	
Dipole (D)		Evaporation time (ether = 1)	
Dielectric constant (20°C)		Evaporation time (BuAc = 1)	0.14
Polarity (water 100)	7.1		

*Thermal information**

Latent heat (cal/mol)	8952
Nett heat of combustion (kcal/gmol)	1179
Specific heat (cal/mol/°C)	48
Critical pressure (MN/m^2)	3.21
Critical temperature (K)	631
Latent heat of fusion (cal/mol)	1680
Van der Waals' volume	5.30
Van der Waals' surface area	4.04
Molar volume	139

*Typical mixture of isomers

Solute	Azeotrope X% w/w	°C	Reference	Solute γ^∞	Reference	Partition coefficient	Reference
Hydrocarbons							
n-Pentane							
n-Hexane							
n-Heptane							
n-Octane							
n-Nonane							
n-Decane							
2,2,4-TMP							
Cyclohexane							
Benzene	None		10882a	1.1	7/322		
Toluene							
Ethylbenzene							
Xylenes							
C_9 Aromatics	–	–	–	–	–	–	
Tetralin							
Alcohols							
Methanol	None		2116				
Ethanol	None		4173				
n-Propanol	None		6529				
i-Propanol	None		6422	2.6	2d/97		
n-Butanol	None		8201				
i-Butanol	None		8382				
s-Butanol				2.2	2f/241		
n-Amyl alc.							
i-Amyl alc.	?	?	9886				
Cyclohexanol	50	156	11750				
1-Octanol	None		14729				
Ethanediol	85	153	4372				
DEG							
1,2-Propanediol							
Glycol ethers							
PGME	30	148	10008				
EGME	15	122	6618	3.3	2d/131		
EEE	10	134	8484				
EGBE	20	160	12252				
Chlorinated							
MDC							
Chloroform							
Carbon tet.							
1,2-EDC							
1,1,1-TCA							
TCE							
Perk.							
MCB	None		10566				

Hydrocarbons

Solute	Azeotrope X% w/w	°C	Reference	Solute γ^{∞}	Reference	Partittion coefficient	Reference
Ketones							
Acetone						0.33	V2/508
MEK							
MIBK							
Cyclohexanone	?	152	11374				
NMP				2.8	3b/463		
Acetophenone							
Ethers							
Diethyl ether							
DIPE							
Dibutyl ether							
MTBE							
1,4-Dioxane							
THF							
Esters							
Me acetate							
Et acetate							
i-Propyl acetate							
n-Butyl acetate							
Cellosolve acetate	25	155	12024				
Miscellaneous							
DMF	Azeo						
DMAc							
DMSO							
Sulfolane							
CS_2							
Acetic acid	None		3225				
Aniline	90	167	11205				
Nitrobenzene							
Morpholine							
Pyridine							
2-Nitropropane							
Acetonitrile							
Furfuraldehyde	40	152	8805				
Phenol	20	158	10971	2.8	2b/385	0.02	V3/301
Water	50	95	767	52.1	1b/286		

Although individual C_9 aromatics can be obtained commercially the mixture used as a solvent usually contains a mixture of up to eight isomers with atmospheric boiling points in the range of 152–176°C. In many cases some isomers form azeotropes with other solvents while others do not. If any azeotropes are reported they are included in the above table.

Tetralin

Alternative names
Tetrahydronaphthalene, tetranap

Reference codes

CAS number	119 64 2	Hazchem code	
UN number		EPA code	

Physical properties

Molecular weight	132	Cubic expansion coeff (per °C x 10^3)	1.0
Empirical formula	$C_{10}H_{12}$	Surface tension (@20°C dyn/cm)	35.5
Boiling point (°C)	205	Absolute viscosity (@25°C cP)	2.0
Freezing point (°C)	–31	Refractive index (25°C)	1.539
Specific gravity (20/4)	0.974		

Fire hazards

Flash point (closed cup °C)	74	Lower explosive limit (ppm)	8000
Autoignition temperature (°C)	225	Upper explosive limit (ppm)	50000
Electrical conductivity			

Health hazards

IDLH (ppm)		Vapour concentration @21°C ppm	526
OES-TWA	25	Vapour density (relative to air)	4.6
OES-STEL		Vapour pressure @21°C mmHg	0.4
Odour threshold (ppm)		POCP	

Aqueous effluent

Solubility in water (25°C %w/w)	
Solubility of water in (25°C %w/w)	
Log_{10} activated carbon partition	4.8
Log_{10} partition in octanol/water (w/w)	
Biological oxygen demand w/w (days)	0 (5)
Theoretical oxygen demand w/w	3.15

Vapour pressure equation constants (Log_{10}, mmHg)

Antoine equation	A	
	B	
	C	
Cox chart	A	7.52287
	B	2019.3

Solvent properties

Solubility parameter		Kauri butanol value	
Dipole (D)	0	Evaporation time (ether = 1)	200
Dielectric constant (20°C)	2.77	Evaporation time (BuAc = 1)	
Polarity (water 100)	9.3		

Thermal information

Latent heat (cal/mol)	10098
Nett heat of combustion (kcal/gmol)	1289
Specific heat (cal/mol/°C)	0.53
Critical pressure (MN/m^2)	3.65
Critical temperature (K)	447
Latent heat of fusion (cal/mol)	2978
Van der Waals' volume	
Van der Waals' surface area	
Molar volume	136

Solute	Azeotrope X% w/w	Azeotrope °C	Azeotrope Reference	Solute γ^{∞}	Reference	Partition coefficient	Reference
Hydrocarbons							
n-Pentane							
n-Hexane				1.9	1x/4/1398		
n-Heptane				1.9	1x/4/1398		
n-Octane							
n-Nonane							
n-Decane							
2,2,4-TMP							
Cyclohexane				1.4	1x/4/1398		
Benzene				1.0	1x/4/1398		
Toluene							
Ethylbenzene							
Xylenes							
C_9 Aromatics							
Tetralin	–		–	–	–	–	–
Alcohols							
Methanol							
Ethanol							
n-Propanol							
i-Propanol							
n-Butanol				2.2	2b/235		
i-Butanol							
s-Butanol							
n-Amyl alc.							
i-Amyl alc.							
Cyclohexanol							
1-Octanol	13	194		2.1	2f/541		
Ethanediol							
DEG							
1,2-Propanediol							
Glycol ethers							
PGME							
EGME							
EEE							
EGBE							
Chlorinated							
MDC							
Chloroform							
Carbon tet.							
1,2-EDC							
1,1,1-TCA							
TCE							
Perk.							
MCB							

Solute	Azeotrope			Solute γ^∞	Reference	Partition coefficient	Reference
	X% w/w	°C	Reference				
Ketones							
Acetone							
MEK							
MIBK							
Cyclohexanone							
NMP							
Acetophenone							
Ethers							
Diethyl ether							
DIPE							
Dibutyl ether							
MTBE							
1,4-Dioxane							
THF							
Esters							
Me acetate							
Et acetate							
i-Propyl acetate							
n-Butyl acetate							
Cellosolve acetate							
Miscellaneous							
DMF							
DMAc							
DMSO							
Sulfolane							
CS_2							
Acetic acid						4.6	V2/310
Aniline							
Nitrobenzene							
Morpholine							
Pyridine							
2-Nitropropane							
Acetonitrile							
Furfuraldehyde							
Phenol							
Water	20	99	806				

Section 2

Alcohols

Methanol

Alternative names
Methyl alcohol, wood alcohol, carbinol, **not** methylated spirit

Reference codes

CAS number	67 56 1	Hazchem code	2PE
UN number	1230	EPA code	U154

Physical properties

Molecular weight	32	Cubic expansion coeff (per °C × 10^3)	1.2
Empirical formula	$C_1H_4O_1$	Surface tension (@20°C dyn/cm)	22.6
Boiling point (°C)	64	Absolute viscosity (@25°C cP)	0.6
Freezing point (°C)	–98	Refractive index (25°C)	1.326
Specific gravity (20/4)	0.792		

Fire hazards

Flash point (closed cup °C)	15	Lower explosive limit (ppm)	60000
Autoignition temperature (°C)	470	Upper explosive limit (ppm)	365000
Electrical conductivity	1.5E–9		

Health hazards

IDLH (ppm)	25000	Vapour concentration @21°C ppm	156000
OES-TWA	200	Vapour density (relative to air)	1.11
OES-STEL	250	Vapour pressure @21°C mmHg	103
Odour threshold (ppm)	6000	POCP	12.3

Aqueous effluent

Solubility in water (25°C %w/w)	Total
Solubility of water in (25°C %w/w)	Total
Log_{10} activated carbon partition	0.86
Log_{10} partition in octanol/water (w/w)	–0.82
Biological oxygen demand w/w (days)	1.12 (5)
Theoretical oxygen demand w/w	1.5

Vapour pressure equation constants (Log_{10}, mmHg)

Antoine equation	A	8.08097
	B	1582.271
	C	239.726
Cox chart	A	8.23606
	B	1579.9

Solvent properties

Solubility parameter	14.5	Kauri butanol value	380
Dipole (D)	1.7	Evaporation time (ether = 1)	6.3
Dielectric constant (20°C)	32.6	Evaporation time (BuAc = 1)	4.1
Polarity (water 100)	76.2		

Thermal information

Latent heat (cal/mol)	8426
Nett heat of combustion (kcal/gmol)	150
Specific heat (cal/mol/°C)	19.5
Critical pressure (MN/m^2)	7.96
Critical temperature (K)	513
Latent heat of fusion (cal/mol)	758
Van der Waals' volume	1.43
Van der Waals' surface area	1.43
Molar volume	40.4

Solute	Azeotrope X% w/w	°C	Reference	Solute γ^{∞}	Reference	Partition coefficient	Reference
Hydrocarbons							
n-Pentane	8	31	2055	11.2	2e/132		
n-Hexane	50	68	2087	16.0	2a/253		
n-Heptane	52	59	2101	22.1	2c/243		
n-Octane	72	63	2113	45.1	2c/249		
n-Nonane	83	64	2120	38.6	2e/191		
n-Decane	None		2126	49.7	2e/193		
2,2,4-TMP	53	59	2114	28.1	2c/250		
Cyclohexane	38	54	2079	18.3	2a/239		
Benzene	39	58	2066	7.4	2a/205		
Toluene	69	64	2098	8.4	2a/268		
Ethylbenzene	None		2106	10.3	2c/245		
Xylenes	None		2108	10.3	2c/247		
C_9 Aromatics	None		2116				
Tetralin							
Alcohols							
Methanol	–		–	–	–	–	–
Ethanol	None		1944	1.0	2a/50		
n-Propanol	None			1.3	2a/122		
i-Propanol	None		1978a	0.9	2a/123		
n-Butanol	None		2015	1.3	2a/169		
i-Butanol	None			1.3	2a/171		
s-Butanol				0.7	2c/128		
n-Amyl alc.	None		2056	1.4	2a/202		
i-Amyl alc.	None			1.4	2a/201		
Cyclohexanol							
1-Octanol							
Ethanediol	None		1945		2a/62		
DEG							
1,2-Propanediol							
Glycol ethers							
PGME							
EGME	None		1979		2c/98		
EEE							
EGBE							
Chlorinated							
MDC	93	38	1544	2.1	2a/24		
Chloroform	87	38	1430	2.7	2a/23		
Carbon tet.	79	56	1090	7.0	2a/1		
1,2-EDC	32	61	1930	5.4	2a/44		
1,1,1-TCA	22	56	1923				
TCE	38	59	1915	8.3	2a/40		
Perk.	64	64	1914	17.9	2a/37		
MCB	None		2063	7.8	2a/204		

Solute	Azeotrope X% w/w	°C	Reference	Solute γ^∞	Reference	Partition coefficient	Reference
Ketones							
Acetone	12	56	1963	1.8	2a/68		
MEK	70	64	1993	2.1	2a/133		
MIBK	None		2084	3.3	2a/248		
Cyclohexanone							
NMP							
Acetophenone							
Ethers							
Diethyl ether	98	62		3.2	2a/170		
DIPE	24	57	2091a	4.3	2a/261		
Dibutyl ether				10.6			
MTBE	10	51	2058	3.1	2c/160		
1,4-Dioxane	None		1998	2.2	2a/148		
THF	31	59	1996	2.2	2a/141		
Esters							
Me acetate	19	54	1967	2.9	2a/92		
Et acetate	46	62	1999	2.8	2a/154		
i-Propyl acetate	80	65	2046				
n-Butyl acetate				5.8	2c/213		
Cellosolve acetate							
Miscellaneous							
DMF	None			0.8	2a/115		
DMAc							
DMSO				0.25	2c/62		
Sulfolane				4.0	2c/125		
CS_2	29	40	1175	12.2	2a/35		
Acetic acid	None		1933	0.9	2a/48		
Aniline							
Nitrobenzene	None		2065				
Morpholine							
Pyridine	None		2024	1.0	2a/183		
2-Nitropropane	None		1977				
Acetonitrile	19	64	1925	2.4	2a/43		
Furfuraldehyde				1.0	2c/140		
Phenol							
Water	None		213	1.5	1/49		

Ethanol

Alternative names
Ethyl alcohol, grain alcohol, methylated spirits, IMS

Reference codes

CAS number	64 17 5	Hazchem code	2SE
UN number	1170	EPA code	U001

Physical properties

Molecular weight	46	Cubic expansion coeff (per °C × 10^3)	1.1
Empirical formula	$C_2H_6O_1$	Surface tension (@20°C dyn/cm)	22.3
Boiling point (°C)	78	Absolute viscosity (@25°C cP)	1.08
Freezing point (°C)	–114	Refractive index (25°C)	1.359
Specific gravity (20/4)	0.789		

Fire hazards

Flash point (closed cup °C)	13	Lower explosive limit (ppm)	33000
Autoignition temperature (°C)	419	Upper explosive limit (ppm)	190000
Electrical conductivity	1.4E–9		

Health hazards

IDLH (ppm)		Vapour concentration @21°C ppm	64000
OES-TWA	1000	Vapour density (relative to air)	1.6
OES-STEL		Vapour pressure @21°C mmHg	45.7
Odour Threshold (ppm)	6000	POCP	27

Aqueous effluent

Solubility in water (25°C %w/w)	Total
Solubility of water in (25°C %w/w)	Total
Log_{10} activated carbon partition	1.35
Log_{10} partition in octanol/water (w/w)	–0.32
Biological oxygen demand w/w (days)	0.92 (5)
Theoretical oxygen demand w/w	2.09

Vapour pressure equation constants (Log_{10}, mmHg)

Antoine equation	A	8.11220
	B	1592.864
	C	226.184
Cox chart	A	8.24183
	B	1651.2

Solvent properties

Solubility parameter	13.4	Kauri butanol value	325
Dipole (D)	1.7	Evaporation time (ether = 1)	8.3
Dielectric constant (20°C)	22.4	Evaporation time (BuAc = 1)	2.4
Polarity (water 100)	65.4		

Thermal information

Latent heat (cal/mol)	9200
Nett heat of combustion (kcal/gmol)	296
Specific heat (cal/mol/°C)	27
Critical pressure (MN/m^2)	6.39
Critical temperature (K)	516
Latent heat of fusion (cal/mol)	1198
Van der Waals' volume	2.11
Van der Waals' surface area	1.97
Molar volume	58.68

Solute	Azeotrope X% w/w	°C	Reference	Solute γ^{∞}	Reference	Partition coefficient	Reference
Hydrocarbons							
n-Pentane	5	34	4062	6.9	2c/375		
n-Hexane	21	59	4106	8.9	2a/353		
n-Heptane	72	98	4139	11.3	2a/498		
n-Octane	78	77	4165	15.1	2c/462		
n-Nonane				24.5	2e/398		
n-Decane				14.5	2a/508		
2,2,4-TMP	40	72	4167	10.8	2a/503		
Cyclohexane	31	65	4087	7.5	2a/430		
Benzene	32	68	4073	4.0	2a/399		
Toluene	63	77	4120	5.9	2a/477		
Ethylbenzene	None		4144	6.4	2c/460		
Xylenes	None		4146	7.7	2a/500		
C_9 Aromatics	None		4175				
Tetralin							
Alcohols							
Methanol	None		1944	1.1	2a/60		
Ethanol	–		–	–	–	–	–
n-Propanol	None		3981	1.1	2a/236		
i-Propanol	None		3980	1.0	2a/341		
n-Butanol	None		4026	1.0	2a/365		
i-Butanol	None		4029a				
s-Butanol	None		4027	1.0	2a/366		
n-Amyl alc.	None		4063	1.1	2a/396		
i-Amyl alc.	None		1066	1.3	2a/395		
Cyclohexanol	None			2.3	2c/421		
1-Octanol	None						
Ethanediol	None		6.5	2c/297			
DEG	None						
1,2-Propanediol	None		1.9	2c/319			
Glycol ethers							
PGME							
EGME	None		3982				
EEE	None		4032		2c/350		
EGBE	None						
Chlorinated							
MDC	2	40	1551	1.7	2c/283		
Chloroform	7	59	1442	2.0	2a/285		
Carbon tet.	16	65	1105	4.7	2a/276		
1,2-EDC	37	70	2964	3.6	2a/299		
1,1,1-TCA							
TCE	28	71	2286	4.7	2a/295		
Perk.	63	77	2162	6.1	2c/285		
MCB	None		4070	5.6	2a/397		

Alcohols

Solute	Azeotrope X% w/w	°C	Reference	Solute γ^∞	Reference	Partition coefficient	Reference
Ketones							
Acetone	None		3965	1.8	2a/321		
MEK	39	74	4005	1.7	2a/343		
MIBK	None		4101	2.1	2c/423		
Cyclohexanone							
NMP							
Acetophenone							
Ethers							
Diethyl ether	None		4029	2.6	2a/375		
DIPE	17	64	4110	4.1	2a/459		
Dibutyl ether	None		4170	5.0	2e/391		
MTBE							
1,4-Dioxane	>98	78	4011	2.2	2a/348		
THF	10		4009	1.5	2a/328		
Esters							
Me acetate	3	57	3969	1.9	2a/335		
Et acetate	26	72	4012	2.4	2a/351		
i-Propyl acetate	52	77	4054	2.1	2a/391		
n-Butyl acetate	None			3.1	2c/426		
Cellosolve acetate							
Miscellaneous							
DMF	None			0.7	2c/371		
DMAc							
DMSO							
Sulfolane				8.1	2c/344		
CS_2	9	43	1189	6.5	2a/281		
Acetic acid				0.7	2c/293		
Aniline	None		4075	4.9	2a/427		
Nitrobenzene	None		4072				
Morpholine				0.6	2c/345		
Pyridine	None		4038	1.0	2c/355		
2-Nitropropane	94	78	3978				
Acetonitrile	56	73	2760	1.9	2a/298		
Furfuraldehyde	None			5.5	2a/383		
Phenol							
Water	96	78	242	2.7	1/165		

n-Propanol

Alternative names
Propan-1-ol, *n*-propyl alcohol, 1-propanol, ethyl carbinol

Reference codes

CAS number	71 23 8	Hazchem code	2SE
UN number	1274	EPA code	

Physical properties

Molecular weight	60	Cubic expansion coeff (per °C × 10^3)	0.96
Empirical formula	$C_3H_8O_1$	Surface tension (@20°C dyn/cm)	23.7
Boiling point (°C)	97	Absolute viscosity (@25°C cP)	1.72
Freezing point (°C)	–127	Refractive index (25°C)	1.383
Specific gravity (20/4)	0.804		

Fire hazards

Flash point (closed cup °C)	25	Lower explosive limit (ppm)	21000
Autoignition temperature (°C)	440	Upper explosive limit (ppm)	135000
Electrical conductivity	9.0E–9		

Health hazards

IDLH (ppm)	4000	Vapour concentration @21°C ppm	18000
OES-TWA	200	Vapour density (relative to air)	2.07
OES-STEL	250	Vapour pressure @21°C mmHg	13.4
Odour threshold (ppm)	45	POCP	45

Aqueous effluent

Solubility in water (25°C %w/w)	Total
Solubility of water in (25°C %w/w)	Total
Log_{10} activated carbon partition	1.67
Log_{10} partition in octanol/water (w/w)	+0.34
Biological oxygen demand w/w (days)	1.5
Theoretical oxygen demand w/w	2.40

Vapour pressure equation constants (Log_{10}, mmHg)

Antoine equation	A	8.37895
	B	1788.020
	C	227.438
Cox chart	A	8.25022
	B	1755.8

Solvent properties

Solubility parameter	11.9	Kauri butanol value	250
Dipole (D)	1.7	Evaporation time (ether = 1)	9.0
Dielectric constant (20°C)	20.1	Evaporation time (BuAc = 1)	1.0
Polarity (water 100)	61.7		

Thermal information

Latent heat (cal/mol)	9780
Nett heat of combustion (kcal/gmol)	438
Specific heat (cal/mol/°C)	34
Critical pressure (MN/m^2)	5.10
Critical temperature (K)	537
Latent heat of fusion (cal/mol)	1240
Van der Waals' volume	2.78
Van der Waals' surface area	2.51
Molar volume	75.14

Solute	Azeotrope X% w/w	°C	Reference	Solute γ^{∞}	Reference	Partition coefficient	Reference
Hydrocarbons							
n-Pentane	None		6484				
n-Hexane	4	66	6505	13.3	2a/584		
n-Heptane	36	88	6514	5.0	2a/596		
n-Octane	70	94	6523	6.6	2a/576		
n-Nonane	98	97		8.3	2e/505		
n-Decane	None			8.7	2a/606		
2,2,4-TMP	41	85	6524	6.5	2e/500		
Cyclohexane	20	74	6495	3.8	2a/579	1.2	CEH
Benzene	17	77	6491	3.3	2a.556		
Toluene	49	93	6512	3.3	2a/592		
Ethylbenzene	91	7	6517	3.8	2a/601		
Xylenes	93	97	6519	4.0	2c/575		
C_9 Aromatics	None		6529				
Tetralin							
Alcohols							
Methanol	None			1.1	2a/122		
Ethanol	None		3981	1.1	2a/336		
n-Propanol	–		–	–	–	–	–
i-Propanol	None			1.1	2a/531		
n-Butanol	None			1.1	2a/539	0.07	CEH
i-Butanol	None		6465	1.1	2a/541		
s-Butanol	None		6463				
n-Amyl alc.	None			1.0	2e/471		
i-Amyl alc.	None		6486	1.0	2a/548		
Cyclohexanol				1.8	2e/414		
1-Octanol							
Ethanediol	None		4195a	6.2	2c/483		
DEG							
1,2-Propanediol				2.9	2c/491		
Glycol ethers							
PGME							
EGME	None			1.3	2c/490		
EEE							
EGBE							
Chlorinated							
MDC	None			2.7	2e/416		
Chloroform	None		1454				
Carbon tet.	8	73	1116	3.3	2a/509		
1,2-EDC	19	81	2971	2.4	2a/520		
1,1,1-TCA	7	73					
TCE	17	82	2296	5.9	2a/518		
Perk.	48	94	2177				
MCB	80	97	6489	3.5	2a/552		

Solute	Azeotrope X% w/w	°C	Reference	Solute γ^∞	Reference	Partition coefficient	Reference
Ketones							
Acetone	None		5320				
MEK	None		6445	1.6	2c/496		
MIBK	35	94					
Cyclohexanone							
NMP				0.2	2e/461		
Acetophenone							
Ethers							
Diethyl ether	None		6464				
DIPE	None			3.6	2a/586		
Dibutyl ether				3.2			
MTBE							
1,4-Dioxane	55	95	6447	1.8	2a/533		
THF	None			1.2	2c/497		
Esters							
Me acetate	None			3.6	2a/530		
Et acetate	None			1.7	2a/536		
i-Propyl acetate							
n-Butyl acetate	40	94		14.5	2e/484		
Cellosolve acetate							
Miscellaneous							
DMF							
DMAc				0.4	2e/454		
DMSO							
Sulfolane							
CS_2	5	46	1209	3.6	2e/417		
Acetic acid				0.9	2a/525		
Aniline							
Nitrobenzene							
Morpholine							
Pyridine	None		6469	0.8	2c/512		
2-Nitropropane	75	96	6271				
Acetonitrile	28	81	2768	3.0	2e/430		
Furfuraldehyde							
Phenol							
Water	71	87	293	3.95	1/301		

i-Propanol

Alternative names
Propan-2-ol, isopropyl alcohol, IPA – avoid confusion with IP acetate

Reference codes

CAS number	67 63 0	Hazchem code	2SE
UN number	1219	EPA code	

Physical properties

Molecular weight	60	Cubic expansion coeff (per °C × 10^3)	1.05
Empirical formula	$C_3H_8O_1$	Surface tension (@20°C dyn/cm)	21.7
Boiling point (°C)	82	Absolute viscosity (@25°C cP)	2.0
Freezing point (°C)	–88	Refractive index (25°C)	1.375
Specific gravity (20/4)	0.786		

Fire hazards

Flash point (closed cup °C)	12	Lower explosive limit (ppm)	23000
Autoignition temperature (°C)	425	Upper explosive limit (ppm)	127000
Electrical conductivity	6.0E–8		

Health hazards

IDLH (ppm)	20000	Vapour concentration @21°C ppm	46000
OES-TWA	400	Vapour density (relative to air)	2.07
OES-STEL	500	Vapour pressure @21°C mmHg	35.1
Odour threshold (ppm)	60	POCP	15

Aqueous effluent

Solubility in water (25°C %w/w)	Total
Solubility of water in (25°C %w/w)	Total
Log_{10} activated carbon partition	1.46
Log_{10} partition in octanol/water (w/w)	+0.26
Biological oxygen demand w/w (days)	1.59
Theoretical oxygen demand w/w	2.40

Vapour pressure equation constants (Log_{10}, mmHg)

Antoine equation	A	8.87829
	B	2010.33
	C	252.636
Cox chart	A	8.24362
	B	1673.2

Solvent properties

Solubility parameter	11.5	Kauri butanol value	230
Dipole (D)	1.66	Evaporation time (ether = 1)	11
Dielectric constant (20°C)	18.3	Evaporation time (BuAc = 1)	1.5
Polarity (water 100)	54.6		

Thermal information

Latent heat (cal/mol)	9540
Nett heat of combustion (kcal/gmol)	433
Specific heat (cal/mol/°C)	37
Critical pressure (MN/m^2)	4.76
Critical temperature (K)	508
Latent heat of fusion (cal/mol)	1282
Van der Waals' volume	2.78
Van der Waals' surface area	2.51
Molar volume	76.92

Solute	Azeotrope X% w/w	°C	Reference	Solute γ^∞	Reference	Partition coefficient	Reference
Hydrocarbons							
n-Pentane	6	35	6370				
n-Hexane	23	63	6390	5.0	2b/97		
n-Heptane	51	76	6399	7.4	2b/113		
n-Octane	84	82	6418	7.8	2b/115		
n-Nonane	Azeo			7.9	2f/95		
n-Decane	None			6.6	2b/118		
2,2,4-TMP	54	77	6419	4.8	2b/116		
Cyclohexane	32	69	6384	4.7	2b/84		
Benzene	33	72	6375	4.0	2b/65		
Toluene	69	81	6397	3.8	2b/108		
Ethylbenzene	None		6402	5.3	2d/95		
Xylenes	None		6404	5.0	2d/96		
C_9 Aromatics	None		6423	5.0	2d/97		
Tetralin							
Alcohols							
Methanol	None		1978a	0.9	2e/123		
Ethanol	None		3980	1.1	2a/341		
n-Propanol	None			1.02	2f/47		
i-Propanol	–		–	–		–	–
n-Butanol	None			1.6	2d/55		
i-Butanol	None			1.0	2d/56		
s-Butanol	None			1.1	2b/62		
n-Amyl alc.	None			0.82	2f/63		
i-Amyl alc.	None						
Cyclohexanol	None						
1-Octanol	None						
Ethanediol	None						
DEG	None						
1,2-Propanediol	None			2.6	2d/47		
Glycol ethers							
PGME							
EGME							
EEE							
EGBE							
Chlorinated							
MDC	None		1561	2.31	2f/36		
Chloroform	4	61	1453	1.6	2d/40		
Carbon tet.	18	69	1115	3.3	2b/36		
1,2-EDC	43	75	2970				
1,1,1-TCA	None		2729				
TCE	30	75	2295	4.0	2d/43		
Perk.	70	82	2176	5.7	2d/42		
MCB	None		6373	4.9	2d/64		

i-Butanol

Alternative names
Isopropyl carbinol, isobutyl alcohol, IBA, 2-me-1-propanol

Reference codes

CAS number	73 83 1	Hazchem code	3Y
UN number	1120	EPA code	U140

Physical properties

Molecular weight	74	Cubic expansion coeff (per °C × 10^3)	0.95
Empirical formula	$C_4H_{10}O_1$	Surface tension (@20°C dyn/cm)	22.8
Boiling point (°C)	108	Absolute Viscosity (@25°C cP)	3.96
Freezing point (°C)	–108	Refractive index (25°C)	1.394
Specific gravity (20/4)	0.802		

Fire hazards

Flash point (closed cup °C)	25	Lower explosive limit (ppm)	16000
Autoignition temperature (°C)	390	Upper explosive limit (ppm)	109000
Electrical conductivity	1.6E–8		

Health hazards

IDLH (ppm)	8000	Vapour concentration @21°C ppm	11500
OES-TWA	50	Vapour density (relative to air)	2.56
OES-STEL	75	Vapour pressure @21°C mmHg	8.6
Odour threshold (ppm)	80	POCP	40

Aqueous effluent

Solubility in water (25°C %w/w)	8.7
Solubility of water in (25°C %w/w)	15.0
Log_{10} activated carbon partition	2.16
Log_{10} partition in octanol/water (w/w)	+0.74
Biological oxygen demand w/w (days)	1.62
Theoretical oxygen demand w/w	2.6

Vapour pressure equation constants (Log_{10}, mmHg)

Antoine equation	A	8.53516
	B	1950.94
	C	237.147
Cox chart	A	8.25506
	B	1816.5

Solvent properties

Solubility parameter	10.7	Kauri butanol value	
Dipole (D)	1.7	Evaporation time (ether = 1)	25
Dielectric constant (20°C)	17.7	Evaporation time (BuAc = 1)	0.6
Polarity (water 100)	55.2		

Thermal information

Latent heat (cal/mol)	10220
Nett heat of combustion (kcal/gmol)	585
Specific heat (cal/mol/°C)	53
Critical pressure (MN/m^2)	4.30
Critical temperature (K)	548
Latent heat of fusion (cal/mol)	
Van der Waals' volume	3.45
Van der Waals' surface area	3.05
Molar volume	92.91

Solute	Azeotrope X% w/w	°C	Reference	Solute γ^{∞}	Reference	Partition coefficient	Reference
Hydrocarbons							
n-Pentane	None		8327				
n-Hexane	2	68	8354	5.3	2f/320		
n-Heptane	27	91	8368	4.0	2f/326		
n-Octane		104	8377				
n-Nonane							
n-Decane							
2,2,4-TMP	27	92	8378				
Cyclohexane	14	78	8338	3.2	2f/317		
Benzene	8	79	8333	3.4	2f/316		
Toluene	45	101	8361	2.7	2b/289		
Ethylbenzene	80	107	8371				
Xylenes	88	107	8373	2.8	2b/292		
C_9 Aromatics							
Tetralin							
Alcohols							
Methanol	None						
Ethanol	None		4029a				
n-Propanol	None		6465	0.9	2f/50	0.24	V2/552
i-Propanol	None			1.0	2d/56		
n-Butanol	None		8103	1.0	2f/153		
i-Butanol	–		–	–	–	–	–
s-Butanol	None			1.0	2f/223		
n-Amyl alc.	None			0.9	2f/311		
i-Amyl alc.	None		8328				
Cyclohexanol							
1-Octanol							
Ethanediol	None		2.3	2f/12			
DEG							
1,2-Propanediol							
Glycol ethers							
PGME							
EGME	None		6548				
EEE	None		8306				
EGBE							
Chlorinated							
MDC							
Chloroform	None		1475				
Carbon tet.	5	76	1137				
1,2-EDC	6	83	2988	2.1	2b/272		
1,1,1-TCA							
TCE	9	85	2309				
Perk.	40	103	2189				
MCB	63	107	8331	2.5	2d/357		

Solute	Azeotrope X% w/w	°C	Reference	Solute γ^∞	Reference	Partition coefficient	Reference
Ketones							
Acetone	None		5344	1.1	2b/140	0.31	V2/469
MEK	None		7357	2.0	2b/143	0.14	V4/237
MIBK	30	114	8152	1.8	2d/193		
Cyclohexanone							
NMP							
Acetophenone							
Ethers							
Diethyl ether	None		8104	1.7	1x/3/1072		
DIPE	None			2.4	2b/202		
Dibutyl ether	83	118	8195	2.4	2d/231		
MTBE							
1,4-Dioxane	None			1.1	2b/147		
THF	None			1.2	2b/146		
Esters							
Me acetate				1.0	2f/137		
Et acetate	None			1.8	2b/148		
i-Propyl acetate	None		8121				
n-Butyl acetate	63	116	8153	1.7	2b/197		
Cellosolve acetate	None		8160				
Miscellaneous							
DMF							
DMAc							
DMSO	None			0.5	2f/131		
Sulfolane							
CS_2	None		1233	2.6	2f/120		
Acetic acid	57	120	3135	1.0	2d/158	0.20	P307
Aniline	None		8138				
Nitrobenzene	None		8135				
Morpholine							
Pyridine	70	119	8109	0.7	2b/166		
2-Nitropropane	48	112	6275				
Acetonitrile	None			4.1	2d/156		
Furfuraldehyde				3.2	2f/155		
Phenol							
Water	58	93	372	5.1	1/407		

n-Butanol

Alternative names
Butyl alcohol, 1-butanol, butanol, propyl carbinol

Reference codes

CAS number	71 36 3	Hazchem code	3Y
UN number	1120	EPA code	U031

Physical properties

Molecular weight	74	Cubic expansion coeff (per °C × 10^3)	0.79
Empirical formula	$C_4H_{10}O_1$	Surface tension (@20°C dyn/cm)	24.6
Boiling point (°C)	118	Absolute viscosity (@25°C cP)	3.0
Freezing point (°C)	–80	Refractive index (25°C)	1.397
Specific gravity (20/4)	0.810		

Fire hazards

Flash point (closed cup °C)	35	Lower explosive limit (ppm)	14000
Autoignition temperature (°C)	360	Upper explosive limit (ppm)	112000
Electrical conductivity	9.1E–9		

Health hazards

IDLH (ppm)	8000	Vapour concentration @21°C ppm	6300
OES-TWA	50	Vapour density (relative to air)	2.55
OES-STEL	50	Vapour pressure @21°C mmHg	4.8
Odour threshold (ppm)	5000	POCP	55

Aqueous effluent

Solubility in water (25°C %w/w)	7.3
Solubility of water in (25°C %w/w)	20.4
Log_{10} activated carbon partition	2.36
Log_{10} partition in octanol/water (w/w)	+0.88
Biological oxygen demand w/w (days)	1.15 (5)
Theoretical oxygen demand w/w	2.21

Vapour pressure equation constants (Log_{10}, mmHg)

Antoine equation	A	7.83800
	B	1558.19
	C	196.881
Cox chart	A	8.25925
	B	1871.7

Solvent properties

Solubility parameter	11.4	Kauri butanol value	225
Dipole (D)	1.66	Evaporation time (ether = 1)	33
Dielectric constant (20°C)	18.2	Evaporation time (BuAc = 1)	0.47
Polarity (water 100)	60.2		

Thermal information

Latent heat (cal/mol)	10434
Nett heat of combustion (kcal/gmol)	586
Specific heat (cal/mol/°C)	41
Critical pressure (MN/m^2)	4.41
Critical temperature (K)	563
Latent heat of fusion (cal/mol)	2215
Van der Waals' volume	3.45
Van der Waals' surface area	3.05
Molar volume	91.97

Solute	Azeotrope X% w/w	°C	Reference	Solute γ^∞	Reference	Partition coefficient	Reference
Hydrocarbons							
n-Pentane	None			4.1	2b/169		
n-Hexane	3	68	8163	2.8	2b/200		
n-Heptane	18	94	8182	4.2	2b/218		
n-Octane	50	110	8194	6.6	1x/3/1075		
n-Nonane	72	116	8203	10.0	2f/209		
n-Decane	8			8.9	2b/236		
2,2,4-TMP							
Cyclohexane	4	80	8146	3.3	2b/188		
Benzene	None		8136	2.1	2f/169		
Toluene	28	105	8170	2.3	2b/207		
Ethylbenzene	67	115	8185	2.7	2b/228		
Xylenes	73	115	8186	2.7	2b/229		
C_9 Aromatics	None		8198				
Tetralin				3.1	2b/235		
Alcohols							
Methanol	None		2015	1.3	2a/169	0.84	V2/99
Ethanol	None		4026	1.0	2a/365	0.34	V4/205
n-Propanol	None			1.0	2a/539	0.12	V4/226
i-Propanol	None			1.0	2d/55	0.24	V2/590
n-Butanol	–		–	–	–	–	–
i-Butanol	None		8103	1.0	2b/161	0.07	V3/110
s-Butanol	None		8102	0.9	2b/154		
n-Amyl alc.	None			1.0	2b/173		
i-Amyl alc.	None			1.1	2b/170		
Cyclohexanol	None			1.2	2b/193		
1-Octanol							
Ethanediol				2.6	2d/6	1.10	V2/420
DEG				1.8	2d/174		
1,2-Propanediol				1.8	2d/137		
Glycol ethers							
PGME							
EGME							
EEE	None		8106				
EGBE	None		8106	1.1	2f/189		
Chlorinated							
MDC				1.9	1x/1/130		
Chloroform	None			1.5	2b/136		
Carbon tet.	3	77	1133	2.4	2b/135		
1,2-EDC	None		2984	2.3	2b/137		
1,1,1-TCA				2.0	2f/123		
TCE	3	87	2306	2.8	2f/121		
Perk.	30	109	2186	3.1	2d/155		
MCB	56	115	8133	2.5	2b/175		

Solute	Azeotrope X% w/w	Azeotrope °C	Azeotrope Reference	Solute γ^∞	Reference	Partition coefficient	Reference
Ketones							
Acetone	None		5319	2.4	2b/43		
MEK	32	78	6335	1.5	2b/54		
MIBK	None		6386	1.7	2b/96		
Cyclohexanone							
NMP							
Acetophenone							
Ethers							
Diethyl ether	None		6351				
DIPE	15	66	6391	2.9	2b/101		
Dibutyl ether				3.9			
MTBE							
1,4-Dioxane	None		6337	1.7	2b/56		
THF	None		6335a	1.4	2b/55		
Esters							
Me acetate	None		5516	2.5	2b/50		
Et acetate	25	75	6338	1.7	2b/59		
i-Propyl acetate	52	80	6363	1.77	2f/59		
n-Butyl acetate	None			2.0	2d/75		
Cellosolve acetate							
Miscellaneous							
DMF	None						
DMAc							
DMSO				2.5	2f/39		
Sulfolane				13.6	2d/53		
CS_2	8	44	1208				
Acetic acid				0.6	2d/44		
Aniline							
Nitrobenzene							
Morpholine							
Pyridine				0.9	2d/57		
2-Nitropropane	96	82	6270				
Acetonitrile	48	75	2767	2.57	2f/40		
Furfuraldehyde							
Phenol							
Water	88	80	292	3.2	1/334		

n-Amyl alcohol

Alternative names
1-Pentanol, pentyl alcohol, butyl carbinol

Reference codes

CAS number	71 41 0	Hazchem code	
UN number	1105	EPA code	

Physical properties

Molecular weight	88	Cubic expansion coeff (per °C × 10^3)	0.92
Empirical formula	$C_5H_{12}O_1$	Surface tension (@20°C dyn/cm)	25.6
Boiling point (°C)	138	Absolute viscosity (@25°C cP)	4.0
Freezing point (°C)	–78	Refractive index (25°C)	1.408
Specific gravity (20/4)	0.815		

Fire hazards

Flash point (closed cup °C)	48	Lower explosive limit (ppm)	11000
Autoignition temperature (°C)	360	Upper explosive limit (ppm)	100000
Electrical conductivity			

Health hazards

IDLH (ppm)		Vapour concentration @21°C ppm	4030
OES-TWA		Vapour density (relative to air)	3.1
OES-STEL	150	Vapour pressure @21°C mmHg	3.0
Odour threshold (ppm)	10	POCP	

Aqueous effluent

Solubility in water (25°C %w/w)	1.7
Solubility of water in (25°C %w/w)	9.2
Log_{10} activated carbon partition	2.74
Log_{10} partition in octanol/water (w/w)	+1.40
Biological oxygen demand w/w (days)	
Theoretical oxygen demand w/w	2.73

Vapour pressure equation constants (Log_{10}, mmHg)

Antoine equation	A	7.3982
	B	1435.57
	C	179.8
Cox chart	A	
	B	

Solvent properties

Solubility parameter		Kauri butanol value	
Dipole (D)	1.7	Evaporation time (ether = 1)	
Dielectric constant (20°C)	13.9	Evaporation time (BuAc = 1)	0.3
Polarity (water 100)	56.8		

Thermal information

Latent heat (cal/mol)	10613
Nett heat of combustion (kcal/gmol)	733
Specific heat (cal/mol/°C)	37
Critical pressure (MN/m^2)	3.84
Critical temperature (K)	586
Latent heat of fusion (cal/mol)	2345
Van der Waals' volume	4.13
Van der Waals' surface area	3.59
Molar volume	108.6

Solute	Azeotrope X% w/w	°C	Reference	Solute γ^{∞}	Reference	Partition coefficient	Reference
Hydrocarbons							
n-Pentane				3.8	1x/3/1154		
n-Hexane	None		9758	3.7	1x/3/1155		
n-Heptane	Azeo			4.8	2f/382		
n-Octane		122	9769	3.7	2f/383		
n-Nonane				4.5	1x/3/1156		
n-Decane							
2,2,4-TMP							
Cyclohexane	None		9752	3.3	1x/3/1155		
Benzene	None		9748	2.6	1x/3/1155		
Toluene	None		9760	3.1	1x/3/1156		
Ethylbenzene	40	130	9765				
Xylenes	42	131	9766				
C_9 Aromatics							
Tetralin							
Alcohols							
Methanol	None		2056	1.3	2a/202	0.14	V2/117
Ethanol	None		4063	1.5	2a/396	0.50	V2/348
n-Propanol	None			1.0	2c/471		
i-Propanol	None			0.9	2f/63		
n-Butanol	None			0.8	2b/173		
i-Butanol				1.0	2f/311		
s-Butanol							
n-Amyl alc.	–		–	–	–	–	–
i-Amyl alc.	None		9746a				
Cyclohexanol							
1-Octanol							
Ethanediol	None			4.4	2d/8	2.7	V2/423
DEG							
1,2-Propanediol				1.7	2d/139		
Glycol ethers							
PGME							
EGME	None		6560				
EEE	None		8419				
EGBE							
Chlorinated							
MDC				1.8	1x/3/1153		
Chloroform				1.3	1x/3/1153		
Carbon tet.	None		1147a	1.7	2f/375		
1,2-EDC				1.2	2f/373		
1,1,1-TCA							
TCE							
Perk.	15	117	2201				
MCB	25	126	9747				

Solute	Azeotrope X% w/w	°C)	Reference	Solute γ^∞	Reference	Partition coefficient	Reference
Ketones							
Acetone							
MEK	None		7358	1.4	2b/239		
MIBK							
Cyclohexanone							
NMP							
Acetophenone				6.0	2b/251		
Ethers							
Diethyl ether							
DIPE							
Dibutyl ether							
MTBE							
1,4-Dioxane	40	99	7520				
THF							
Esters							
Me acetate							
Et acetate	None		7568				
i-Propyl acetate							
n-Butyl acetate	None		8237				
Cellosolve acetate							
Miscellaneous							
DMF							
DMAc							
DMSO							
Sulfolane							
CS_2				4.8	2f/218		
Acetic acid				0.9	2f/221	0.20	P309
Aniline				3.8	2b/265		
Nitrobenzene	None		8231	12.5	2f/226		
Morpholine							
Pyridine	None		8217	0.8	2f/224		
2-Nitropropane	82	99	6276				
Acetonitrile				2.3	2b/241		
Furfuraldehyde							
Phenol							
Water	73	87	373	7.3	1/420		

s-Butanol

Alternative names
2-Butanol, methyl ethyl carbinol, 2-hydroxy-butane

Reference codes

CAS number	78 92 2	Hazchem code	3Y
UN number	1120	EPA code	

Physical properties

Molecular weight	74	Cubic expansion coeff (per °C × 10^3)	0.91
Empirical formula	$C_4H_{10}O_1$	Surface tension (@20°C dyn/cm)	23.0
Boiling point (°C)	99.5	Absolute viscosity (@25°C cP)	3.7
Freezing point (°C)	−115	Refractive index (25°C)	1.395
Specific gravity (20/4)	0.807		

Fire hazards

Flash point (closed cup °C)	21	Lower explosive limit (ppm)	17000
Autoignition temperature (°C)	405	Upper explosive limit (ppm)	98000
Electrical conductivity	<1.0E−7		

Health hazards

IDLH (ppm)	10,000	Vapour concentration @21°C ppm	17600
OES-TWA	100	Vapour density (relative to air)	2.56
OES-STEL	150	Vapour pressure @21°C mmHg	13.2
Odour threshold (ppm)	75	POCP	55

Aqueous effluent

Solubility in water (25°C %w/w)	19.8
Solubility of water in (25°C %w/w)	65.1
Log_{10} activated carbon partition	
Log_{10} partition in octanol/water (w/w)	+0.61
Biological oxygen demand w/w (days)	1.87
Theoretical oxygen demand w/w	2.59

Vapour pressure equation constants (Log_{10}, mmHg)

Antoine equation	A	7.47429
	B	1314.188
	C	186.500
Cox chart	A	8.25102
	B	1766.8

Solvent properties

Solubility parameter	10.8	Kauri butanol value	195
Dipole (D)	1.7	Evaporation time (ether = 1)	13.0
Dielectric constant (20°C)	16.56	Evaporation time (BuAc = 1)	0.9
Polarity (water 100)	50.6		

Thermal information

Latent heat (cal/mol)	9916
Nett heat of combustion (kcal/gmol)	583
Specific heat (cal/mol/°C)	40
Critical pressure (MN/m^2)	4.20
Critical temperature (K)	536
Latent heat of fusion (cal/mol)	
Van der Waals' volume	
Van der Waals' surface area	
Molar volume	91.70

Solute	Azeotrope X% w/w	°C	Reference	Solute γ^∞	Reference	Partition coefficient	Reference
Hydrocarbons							
n-Pentane	None		8228				
n-Hexane	8	67	8242	4.3	2b/250		
n-Heptane	37	88	8248	3.8	2f/239		
n-Octane							
n-Nonane							
n-Decane				2.6	2b/285		
2,2,4-TMP	34	88	8254	4.1	2b/284		
Cyclohexane	18	76	8234	1.6	2f/234		
Benzene	15	79	8232	1.8	2f/227		
Toluene	55	95	8246	2.1	2b/276		
Ethylbenzene	None		8251				
Xylenes	None		8252	2.7	2b/282		
C_9 Aromatics				3.4	2f/241		
Tetralin							
Alcohols							
Methanol	None			0.9	2c/128		
Ethanol	None		4027	1.2	2a/366		
n-Propanol	None		6463				
i-Propanol	None			1.1	2b/62		
n-Butanol	None		8102	0.9	2b/154		
i-Butanol	None			1.0	2f/223		
s-Butanol	–		–	–	–	–	–
n-Amyl alc.							
i-Amyl alc.							
Cyclohexanol							
1-Octanol							
Ethanediol							
DEG							
1,2-Propanediol							
Glycol ethers							
PGME							
EGME	None		6547				
EEE							
EGBE							
Chlorinated							
MDC							
Chloroform	None		1472				
Carbon tet.	8	74	1134	2.2	2f/217		
1,2-EDC	12	82	2985	2.2	2f/220		
1,1,1-TCA							
TCE	15	84	2307	2.1	2f/219		
Perk.	57	97	2187	4.0	2b/240		
MCB	None		8230	2.7	2b/258		

Solute	Azeotrope X% w/w	°C	Reference	Solute γ^∞	Reference	Partition coefficient	Reference
Ketones							
Acetone	None		5347	1.4	2f/304		
MEK	None		7360				
MIBK	91	108	8343				
Cyclohexanone							
NMP							
Acetophenone							
Ethers							
Diethyl ether							
DIPE							
Dibutyl ether	None		8379				
MTBE							
1,4-Dioxane	4	101	7522	1.5	2b/278		
THF							
Esters							
Me acetate							
Et acetate	None		7570			0.03	P864
i-Propyl acetate	None		8318				
n-Butyl acetate	None		8345				
Cellosolve acetate	None		8352				
Miscellaneous							
DMF							
DMAc							
DMSO	None			0.4	2b/275		
Sulfolane							
CS_2	None		1236				
Acetic acid	Azeo			1.0	2f/302	0.21	P308
Aniline							
Nitrobenzene							
Morpholine							
Pyridine	None		8307	0.7	2f/307	0.03	P1107
2-Nitropropane	67	105	6278				
Acetonitrile	None		2779	3.4	2f/300		
Furfuraldehyde							
Phenol							
Water	67	89	376	4.8	1/440		

Cyclohexanol

Alternative names
Hexalin, cyclohexyl alcohol

Reference codes			
CAS number	108 93 0	Hazchem code	
UN number		EPA code	
Physical properties			
Molecular weight	100	Cubic expansion coeff (per °C × 10^3)	0.82
Empirical formula	$C_6H_{12}O_1$	Surface tension (@20°C dyn/cm)	32
Boiling point (°C)	161	Absolute viscosity (@25°C cP)	52.7
Freezing point (°C)	+25	Refractive index (25°C)	1.465
Specific gravity (20/4)	0.949		
Fire hazards			
Flash point (closed cup °C)	68	Lower explosive limit (ppm)	20000
Autoignition temperature (°C)	300	Upper explosive limit (ppm)	
Electrical conductivity			
Health hazards			
IDLH (ppm)	3500	Vapour concentration @21°C ppm	1500
OES-TWA	50	Vapour density (relative to air)	3.45
OES-STEL		Vapour pressure @21°C mmHg	1.14
Odour threshold (ppm)	1	POCP	

Aqueous effluent		
Solubility in water (25°C %w/w)		4.3
Solubility of water in (25°C %w/w)		11.8
Log_{10} activated carbon partition		
Log_{10} partition in octanol/water (w/w)		+1.23
Biological oxygen demand w/w (days)		0.08 (5)
Theoretical oxygen demand w/w		2.83
Vapour pressure equation constants (Log_{10}, mmHg)		
Antoine equation	A	8.35237
	B	2258.560
	C	251.624
Cox chart	A	8.27876
	B	2110.6

Solvent properties			
Solubility parameter	11.4	Kauri butanol value	
Dipole (D)	1.8	Evaporation time (ether = 1)	150
Dielectric constant (20°C)	15.0	Evaporation time (BuAc = 1)	0.08
Polarity (water 100)	50.0		

Thermal information	
Latent heat (cal/mol)	10900
Nett heat of combustion (kcal/gmol)	892
Specific heat (cal/mol/°C)	50
Critical pressure (MN/m^2)	3.7
Critical temperature (K)	625
Latent heat of fusion (cal/mol)	419
Van der Waals' volume	4.35
Van der Waals' surface area	3.51
Molar volume	103.43

Solute	Azeotrope X% w/w	°C	Reference	Solute γ^{∞}	Reference	Partition coefficient	Reference
Hydrocarbons							
n-Pentane				5.5	1x/3/1232		
n-Hexane	None			5.7	1x/1/272		
n-Heptane	None		11727	6.8	2f/419		
n-Octane							
n-Nonane							
n-Decane							
2,2,4-TMP							
Cyclohexane	None		11684	3.4	1x/1/272		
Benzene	None		10856	2.7	1x/1/272		
Toluene	None		11720				
Ethylbenzene							
Xylenes	10	140	11730		2f/536		
C_9 Aromatics	40		11750				
Tetralin							
Alcohols							
Methanol							
Ethanol	None			1.2	2c/421		
n-Propanol	None			3.9	2e/414		
i-Propanol	None						
n-Butanol	None			1.0	2b/193		
i-Butanol							
s-Butanol							
n-Amyl alc.							
i-Amyl alc.							
Cyclohexanol	–		–	–	–	–	–
1-Octanol							
Ethanediol	None		4257	2.3	2d/14		
DEG							
1,2-Propanediol							
Glycol ethers							
PGME	None		9967				
EGME							
EEE							
EGBE	None		11712			0.02	P3974
Chlorinated							
MDC							
Chloroform							
Carbon tet.							
1,2-EDC							
1,1,1-TCA							
TCE							
Perk.							
MCB	None			1.6	2b/393		

Alcohols

Solute	Azeotrope			Solute γ^∞	Reference	Partition coefficient	Reference
	X% w/w	°C	Reference				
Ketones							
Acetone							
MEK				1.8	1x/3/1151		
MIBK	None		9836				
Cyclohexanone							
NMP							
Acetophenone							
Ethers							
Diethyl ether							
DIPE							
Dibutyl ether	65	130	9883				
MTBE							
1,4-Dioxane				2.1	1x/3/1151		
THF							
Esters							
Me acetate							
Et acetate							
i-Propyl acetate							
n-Butyl acetate							
Cellosolve acetate	None		9842				
Miscellaneous							
DMF							
DMAc							
DMSO							
Sulfolane							
CS_2							
Acetic acid	84	133	3156a				
Aniline	None		9824				
Nitrobenzene							
Morpholine							
Pyridine	None		8838				
2-Nitropropane							
Acetonitrile							
Furfuraldehyde						0.07	V3/185
Phenol	None		9822				
Water	50	95	468	1.9	1a/382		

i-Amyl alcohol

Alternative names

Fusel oil, 3-methyl-1-butanol, isopentyl alcohol, isobutyl carbinol

Reference codes

CAS number	123 51 3	Hazchem code	
UN number		EPA code	

Physical properties

Molecular weight	88	Cubic expansion coeff (per °C × 10^3)	1.12
Empirical formula	$C_5H_{12}O_1$	Surface tension (@20°C dyn/cm)	23.8
Boiling point (°C)	130	Absolute viscosity (@25°C cP)	4.2
Freezing point (°C)	–134	Refractive index (25°C)	1.4014
Specific gravity (20/4)	0.810		

Fire hazards

Flash point (closed cup °C)	46	Lower explosive limit (ppm)	12000
Autoignition temperature (°C)	365	Upper explosive limit (ppm)	90000
Electrical conductivity			

Health hazards

IDLH (ppm)	8000	Vapour concentration @21°C ppm	3140
OES-TWA	100	Vapour density (relative to air)	3.06
OES-STEL	125	Vapour pressure @21°C mmHg	2.4
Odour threshold (ppm)	15	POCP	

Aqueous effluent

Solubility in water (25°C %w/w)	2.75
Solubility of water in (25°C %w/w)	
Log_{10} activated carbon partition	
Log_{10} partition in octanol/water (w/w)	+1.16
Biological oxygen demand w/w (days)	
Theoretical oxygen demand w/w	2.73

Vapour pressure equation constants (Log_{10}, mmHg)

Antoine equation	A	7.382
	B	1373.8
	C	174.3
Cox chart	A	
	B	

Solvent properties

Solubility parameter		Kauri butanol value	
Dipole (D)	1.8	Evaporation time (ether = 1)	62
Dielectric constant (20°C)	15.2	Evaporation time (BuAc = 1)	
Polarity (water 100)	56.5		

Thermal information

Latent heat (cal/mol)	10542
Nett heat of combustion (kcal/gmol)	792
Specific heat (cal/mol/°C)	47.3
Critical pressure (MN/m^2)	
Critical temperature (K)	583
Latent heat of fusion (cal/mol)	
Van der Waals' volume	4.13
Van der Waals' surface area	3.59
Molar volume	109.2

Solute	Azeotrope X% w/w	Azeotrope °C	Azeotrope Reference	Solute γ^{∞}	Reference	Partition coefficient	Reference
Hydrocarbons							
n-Pentane							
n-Hexane	None		9845				
n-Heptane	7	98	9869				
n-Octane	35	120	9881	4.8	1x/3/1151		
n-Nonane							
n-Decane							
2,2,4-TMP	5	99	9882				
Cyclohexane	None		9834				
Benzene	None		9821				
Toluene	10	110	9852	3.0	1x/3/1151		
Ethylbenzene	49	126	9871				
Xylenes	52	126	9873				
C_9 Aromatics	None		9887				
Tetralin							
Alcohols							
Methanol	None			1.2	2a/201		
Ethanol	None		4066	1.2	2a/395	0.50	V2/347
n-Propanol	None		6486	1.0	2a/548	0.12	V2/566
i-Propanol						0.07	V2/592
n-Butanol	None			1.0	2b/170		
i-Butanol	None		8328			0.08	V3/139
s-Butanol							
n-Amyl alc.							
i-Amyl alc.	–		–	–	–	–	–
Cyclohexanol							
1-Octanol							
Ethanediol	None		4221	4.1	2d/7		
DEG							
1,2-Propanediol	None			1.7	2d/138		
Glycol ethers							
PGME	None		9816				
EGME	None		6562				
EEE	None		8420				
EGBE							
Chlorinated							
MDC							
Chloroform							
Carbon tet.	None		1149				
1,2-EDC	None		2995				
1,1,1-TCA							
TCE	None		2321				
Perk.	19	116	2203				
MCB	34	124	9819				

Solute	Azeotrope X% w/w	°C	Reference	Solute γ^∞	Reference	Partition coefficient	Reference
Ketones							
Acetone				2.1	1x/3/1154		
MEK				1.8	1x/3/1154		
MIBK	Azeo			1.7	2f/380		
Cyclohexanone							
NMP							
Acetophenone							
Ethers							
Diethyl ether				1.7	1x/3/1154		
DIPE							
Dibutyl ether	50	135	9770				
MTBE							
1,4-Dioxane							
THF							
Esters							
Me acetate							
Et acetate				2.1	1x/3/1154		
i-Propyl acetate							
n-Butyl acetate	None		9754				
Cellosolve acetate							
Miscellaneous							
DMF							
DMAc							
DMSO							
Sulfolane							
CS_2	None		1257	2.0	2f/371	0.22	P313
Acetic acid							
Aniline							
Nitrobenzene							
Morpholine							
Pyridine	None		8836				
2-Nitropropane	15	120	6280				
Acetonitrile							
Furfuraldehyde						0.05	V3/186
Phenol	None		9749			<0.01	P1649
Water	46	96	464	3.4	1a/383		

Ethanediol

Alternative names
Glycol, monoethyleneglycol, 1,2-dihydroxyethane, MEG, **not** ethyl glycol

Reference codes

CAS number	107 21 1	Hazchem code	
UN number		EPA code	

Physical properties

Molecular weight	62	Cubic expansion coeff (per °C × 10^3)	0.64
Empirical formula	$C_2H_6O_2$	Surface tension (@20°C dyn/cm)	46.5
Boiling point (°C)	198	Absolute viscosity (@25°C cP)	20
Freezing point (°C)	–13	Refractive index (25°C)	1.429
Specific gravity (20/4)	1.115		

Fire hazards

Flash point (closed cup °C)	111	Lower explosive limit (ppm)	32000
Autoignition temperature (°C)	413	Upper explosive limit (ppm	
Electrical conductivity	1.2E–6		

Health hazards

IDLH (ppm)		Vapour concentration @21°C ppm	153
OES-TWA	60	Vapour density (relative to air)	2.15
OES-STEL	125	Vapour pressure @21°C mmHg	0.12
Odour threshold (ppm)		POCP	

Aqueous effluent

Solubility in water (25°C %w/w)	Total
Solubility of water in (25°C %w/w)	Total
Log_{10} activated carbon partition	1.16
Log_{10} partition in octanol/water (w/w)	–1.93
Biological oxygen demand w/w (days)	0.16 (5)
Theoretical oxygen demand w/w	1.29

Vapour pressure equation constants (Log_{10}, mmHg)

Antoine equation	A	8.09083
	B	2088.936
	C	203.454
Cox chart	A	
	B	

Solvent properties

Solubility parameter	14.6	Kauri butanol value	
Dipole (D)	2.31	Evaporation time (ether = 1)	1550
Dielectric constant (20°C)	37.7	Evaporation time (BuAc = 1)	
Polarity (water 100)	79.0		

Thermal information

Latent heat (cal/mol)	12524
Nett heat of combustion (kcal/gmol)	250
Specific heat (cal/mol/°C)	35
Critical pressure (MN/m^2)	7.7
Critical temperature (K)	647
Latent heat of fusion (cal/mol)	2682
Van der Waals' volume	2.41
Van der Waals' surface area	2.25
Molar volume	55.92

Solute	Azeotrope X% w/w	°C	Reference	Solute γ^{∞}	Reference	Partition coefficient	Reference
Hydrocarbons							
n-Pentane				229.0	1x/1/33		
n-Hexane	None		4265	384.0	1x/1/35		
n-Heptane	3	98	4312	806.0	1x/1/35		
n-Octane	11	124	4353	1170.0	1x/1/36		
n-Nonane				1710.0	1x/1/37		
n-Decane	23	161	4434	1970.0	1x/1/37		
2,2,4-TMP							
Cyclohexane	None		4255	148.0	1x/1/34		
Benzene	None		4239	45.7	1x/1/34		
Toluene	6	110	4285	76.9	1x/1/35		
Ethylbenzene	13	133	4321	86.7	1x/1/36		
Xylenes	16	140	4323	110.0	1x/1/36		
C_9 Aromatics		160	4372	200	1x/1/37		
Tetralin							
Alcohols							
Methanol	None		1946		2a/62		
Ethanol	None			2.1	2c/297		
n-Propanol	None		4195a	4.4	2c/483		
i-Propanol							
n-Butanol	None		4213	6.6	2d/6		
i-Butanol				2.5	2f/12		
s-Butanol							
n-Amyl alc.	None			4.1	2d/8		
i-Amyl alc.	None		4221	3.3	2d/7		
Cyclohexanol	None		4257	2.3	2d/14		
1-Octanol	36	184	4356				
Ethanediol	–		–	–	–	–	–
DEG				1.9	2f/13		
1,2-Propanediol				0.9	2b/12		
Glycol ethers							
PGME	None		4222				
EGME							
EEE							
EGBE	None		4268				
Chlorinated							
MDC							
Chloroform							
Carbon tet.							
1,2-EDC							
1,1,1-TCA							
TCE	None		2287				
Perk.	6	119	2163				
MCB	6	130	4233				

Alcohols

Solute	Azeotrope X w/w	°C	Reference	Solute γ^∞	Reference	Partition coefficient	Reference
Ketones							
Acetone				2.6	1x/1/401	0.24	P491
MEK				2.0	1x/1/401	0.07	P852
MIBK							
Cyclohexanone						0.02	P1839
NMP							
Acetophenone	87	195	13909			<0.01	P2743
Ethers							
Diethyl ether						0.04	P986
DIPE							
Dibutyl ether							
MTBE							
1,4-Dioxane				1.2	1x/3/1371		
THF							
Esters							
Me acetate						0.09	P515
Et acetate				2.4	1x/1/401		
i-Propyl acetate							
n-Butyl acetate							
Cellosolve acetate							
Miscellaneous							
DMF				1.23	2f/527		
DMAc						0.82	P931
DMSO						14.8	P386
Sulfolane							
CS_2				1.8	1x/1/401		
Acetic acid						0.24	P282
Aniline	17	184	11200			0.02	P1710
Nitrobenzene	None		10754			<0.01	P1438
Morpholine						1.66	P932
Pyridine						0.07	V3/234
2-Nitropropane							
Acetonitrile				7.5	1x/1/401	0.30	P260
Furfuraldehyde	None		8789				
Phenol	87	195	10962			<0.01	P1617
Water	10	99	741				

n-Octanol

Alternative names
Octyl alcohol, heptyl carbinol

Reference codes

CAS number	111 87 5	Hazchem code	
UN number		EPA code	

Physical properties

Molecular weight	130	Cubic expansion coeff (per °C × 10^3)	0.99
Empirical formula	$C_8H_{18}O_1$	Surface tension (@20°C dyn/cm)	27.5
Boiling point (°C)	194	Absolute viscosity (@25°C cP)	7.5
Freezing point (°C)	–16	Refractive index (25°C)	1.427
Specific gravity (20/4)	0.827		

Fire hazards

Flash point (closed cup °C)	81	Lower explosive limit (ppm)	30000
Autoignition temperature (°C)		Upper explosive limit (ppm)	
Electrical conductivity	1.4E–7		

Health hazards

IDLH (ppm)		Vapour concentration @21°C ppm	186
OES-TWA		Vapour density (relative to air)	4.5
OES-STEL		Vapour pressure @21°C mmHg	0.14
Odour threshold (ppm)	0.5	POCP	

Aqueous effluent

Solubility in water (25°C %w/w)	0.6
Solubility of water in (25°C %w/w)	
Log_{10} activated carbon partition	
Log_{10} partition in octanol/water (w/w)	3.15
Biological oxygen demand w/w (days)	2.95 (5)
Theoretical oxygen demand w/w	2.95

Vapour pressure equation constants (Log_{10}, mmHg)

Antoine equation	A	6.5422
	B	1139.45
	C	115.9
Cox chart	A	8.29377
	B	2295.1

Solvent properties

Solubility parameter	10.4	Kauri butanol value	
Dipole (D)	1.9	Evaporation time (ether = 1)	
Dielectric constant (20°C)		Evaporation time (BuAc = 1)	
Polarity (water 100)	54.3		

Thermal information

Latent heat (cal/mol)	12675
Nett heat of combustion (kcal/gmol)	1167
Specific heat (cal/mol/°C)	65
Critical pressure (MN/m^2)	2.69
Critical temperature (K)	659
Latent heat of fusion (cal/mol)	
Van der Waals' volume	6.15
Van der Waals' surface area	5.21
Molar volume	158.4

Solute	Azeotrope X% w/w	°C	Reference	Solute γ^∞	Reference	Partition coefficient	Reference
Hydrocarbons							
n-Pentane				2.6	1x/1/401		
n-Hexane				2.8	1x/1/402		
n-Heptane				3.1	1x/1/402		
n-Octane				3.4	1x/1/402		
n-Nonane							
n-Decane				2.8	2f/542		
2,2,4-TMP							
Cyclohexane							
Benzene				2.1	1x/1/401	<0.01	P1483
Toluene				2.1	1x/1/402	<0.01	P2286
Ethylbenzene				2.5	1x/1/402	<0.01	P2874
Xylenes	None		14113	2.4	1x/1/402	<0.01	P2876
C_9 Aromatics	None		14728	2.7	1x/1/402		
Tetralin	87	194		1.4	2f541		
Alcohols							
Methanol						0.76	P155
Ethanol				1.1	1x/3/1371	0.29	P367
n-Propanol						0.06	P636
i-Propanol							
n-Butanol						0.02	P947
i-Butanol						0.03	P964
s-Butanol						0.03	P973
n-Amyl alc.						<0.01	P1266
i-Amyl alc.						<0.01	P1273
Cyclohexanol						<0.01	P1894
1-Octanol	–		–	–	–	–	–
Ethanediol	63	184	4356				
DEG							
1,2-Propanediol							
Glycol ethers							
PGME							
EGME							
EEE						0.48	P1000
EGBE						0.01	P3976
Chlorinated							
MDC				1.6	1x/1/401		
Chloroform				1.0	1x/1/401	<0.01	P87
Carbon tet.				1.7	1x/1/401		
1,2-EDC							
1,1,1-TCA							
TCE							
Perk.							
MCB						<0.01	P1377

Solute	Azeotrope X% w/w	Azeotrope °C	Azeotrope Reference	Solute γ^{∞}	Reference	Part coefficient	Reference
Ketones							
Acetone	None			3.8	2d/510		
MEK							
MIBK							
Cyclohexanone	None		11357		2b/395		
NMP				0.5	2f/411		
Acetophenone							
Ethers							
Diethyl ether							
DIPE							
Dibutyl ether	None		11745				
MTBE							
1,4-Dioxane	None		7543				
THF							
Esters							
Me acetate							
Et acetate	None			1.4	2d/511		
i-Propyl acetate							
n-Butyl acetate	None			1.3	2f/417		
Cellosolve acetate							
Miscellaneous							
DMF							
DMAc							
DMSO							
Sulfolane							
CS_2							
Acetic acid						0.24	V/260
Aniline	None		11149	1.4	2f/416		
Nitrobenzene							
Morpholine							
Pyridine						0.05	V3/234
2-Nitropropane							
Acetonitrile							
Furfuraldehyde	95	156	8764				
Phenol	13	183	10895	2.8	2b/385		
Water	30	98	528	4.5	1/514		

Solute	Azeotrope X% w/w	°C	Reference	Solute γ^∞	Reference	Partition coefficient	Reference
Ketones							
Acetone				6.2	1x/3/971		
MEK				8.4	1x/3/971		
MIBK							
Cyclohexanone							
NMP				0.44	2f/17		
Acetophenone	52	186	4316				
Ethers							
Diethyl ether							
DIPE							
Dibutyl ether	10	140	4354				
MTBE							
1,4-Dioxane	None		4206				
THF	None		4204a	3.60	2d/3		
Esters							
Me acetate							
Et acetate							
i-Propyl acetate							
n-Butyl acetate	None		4258				
Cellosolve acetate	None		4261	6.9	2d/15		
Miscellaneous							
DMF				1.5	2b/8		
DMAc							
DMSO				0.5	2b/7		
Sulfolane							
CS_2							
Acetic acid							
Aniline	24	181	4242	3.9	2b/16		
Nitrobenzene	59	186	4238				
Morpholine							
Pyridine	None		4215				
2-Nitropropane							
Acetonitrile				5.1	2f/1		
Furfuraldehyde	None		4214				
Phenol	78	199	4240	1.1	2d/11		
Water	None		244	0.72	1a/173		

Diethylene glycol

Alternative names
DEG, 2,2-oxydiethanol

Reference codes

CAS number	111 46 6	Hazchem code	
UN number		EPA code	

Physical properties

Molecular weight	106	Cubic expansion coeff (per °C × 10^3)	0.63
Empirical formula	$C_4H_{10}O_3$	Surface tension (@20°C dyn/cm)	48.5
Boiling point (°C)	245	Absolute viscosity (@25°C cP)	34
Freezing point (°C)	–8	Refractive index (25°C)	1.445
Specific gravity (20/4)	1.118		

Fire hazards

Flash point (closed cup °C)	124	Lower explosive limit (ppm)	16000
Autoignition temperature (°C)	229	Upper explosive limit (ppm)	108000
Electrical conductivity	6E–7		

Health hazards

IDLH (ppm)		Vapour concentration @21°C ppm	25
OES-TWA		Vapour density (relative to air)	3.68
OES-STEL		Vapour pressure @21°C mmHg	0.019
Odour threshold (ppm)		POCP	

Aqueous effluent

Solubility in water (25°C %w/w)	Total
Solubility of water in (25°C %w/w)	Total
Log_{10} activated carbon partition	1.86
Log_{10} partition in octanol/water (w/w)	–1.98
Biological oxygen demand w/w (days)	0.06 (5)
Theoretical oxygen demand w/w	1.51

Vapour pressure equation constants (Log_{10}, mmHg)

Antoine equation	A	12.83
	B	7046.4
	C	463.2
Cox chart	A	
	B	

Solvent properties

Solubility parameter		Kauri butanol value	
Dipole (D)	2.31	Evaporation time (ether = 1)	
Dielectric constant (20°C)	31.7	Evaporation time (BuAc = 1)	
Polarity (water 100)	71.3		

Thermal information

Latent heat (cal/mol)	15900
Nett heat of combustion (kcal/gmol)	567
Specific heat (cal/mol/°C)	58.4
Critical pressure (MN/m^2)	4.7
Critical temperature (K)	680
Latent heat of fusion (cal/mol)	
Van der Waals' volume	4.00
Van der Waals' surface area	3.57
Molar volume	94.8

Solute	Azeotrope X% w/w	°C	Reference	Solute γ^{∞}	Reference	Partition coefficient	Reference
Hydrocarbons							
n-Pentane				64.1	1x/1/139		
n-Hexane				91.5	1x/1/141		
n-Heptane				95.6	1x/1/142		
n-Octane				139.8	1x/1/144		
n-Nonane				200.3	1x/1/144		
n-Decane				287.1	1x/1/145		
2,2,4-TMP				195.0	1x/1/144		
Cyclohexane				32.1	1x/1/140		
Benzene	None		8520	5.8	1x/1/139		
Toluene	None		8531	12.3	2f/341		
Ethylbenzene	Azeo		8545	14.9	1x/1/143		
Xylenes	Azeo		8547	17.0	1x/1/143		
C_9 Aromatics				25.0	1x/1/144		
Tetralin							
Alcohols							
Methanol				0.92	1x/3/1079		
Ethanol				1.3	1x/3/1079		
n-Propanol				1.5	1x/3/1079		
i-Propanol				1.4	1x/3/1080		
n-Butanol				1.9	1x/3/1080		
i-Butanol				1.7	1x/3/1080		
s-Butanol							
n-Amyl alc.				2.5	1x/3/1080		
i-Amyl alc.							
Cyclohexanol							
1-Octanol							
Ethanediol							
DEG	–		–	–	–	–	–
1,2-Propanediol							
Glycol ethers							
PGME							
EGME							
EEE							
EGBE							
Chlorinated							
MDC							
Chloroform							
Carbon tet.							
1,2-EDC							
1,1,1-TCA							
TCE							
Perk.							
MCB							

Alcohols

Solute	Azeotrope X% w/w	°C	Reference	Solute γ^∞	Reference	Partition coefficient	Reference
Ketones							
Acetone				3.1	1x/3/1079		
MEK				3.7	1x/3/1080		
MIBK				8.0	1x/3/1081		
Cyclohexanone							
NMP							
Acetophenone	None		8538				
Ethers							
Diethyl ether							
DIPE							
Dibutyl ether							
MTBE							
1,4-Dioxane				2.5	1x/3/1080		
THF							
Esters							
Me acetate				3.4	1x/3/1079		
Et acetate				5.3	1x/3/1080		
i-Propyl acetate							
n-Butyl acetate							
Cellosolve acetate							
Miscellaneous							
DMF							
DMAc							
DMSO							
Sulfolane							
CS_2							
Acetic acid							
Aniline							
Nitrobenzene	10	210	8518				
Morpholine							
Pyridine							
2-Nitropropane							
Acetonitrile							
Furfuraldehyde							
Phenol				0.6	2f/339		
Water	None		386	2.3	1a/353		

1,2-Propanediol

Alternative names
Propylene glycol, **not** propyl glycol

Reference codes

CAS number	57 55 6	Hazchem code	
UN number		EPA code	

Physical properties

Molecular weight	76	Cubic expansion coeff (per °C × 10^3)	0.72
Empirical formula	$C_3H_8O_2$	Surface tension (@20°C dyn/cm)	72
Boiling point (°C)	187	Absolute viscosity (@25°C cP)	54
Freezing point (°C)	–60	Refractive index (25°C)	1.431
Specific gravity (20/4)	1.0362		

Fire hazards

Flash point (closed cup °C)	99	Lower explosive limit (ppm)	26000
Autoignition temperature (°C)	421	Upper explosive limit (ppm)	125000
Electrical conductivity	6.0E–7		

Health hazards

IDLH (ppm)		Vapour concentration @21°C ppm	212
OES-TWA	150	Vapour density (relative to air)	2.52
OES-STEL		Vapour pressure @21°C mmHg	0.16
Odour threshold (ppm)		POCP	

Aqueous effluent

Solubility in water (25°C %w/w)	Total
Solubility of water in (25°C %w/w)	Total
Log_{10} activated carbon partition	1.43
Log_{10} partition in octanol/water (w/w)	–1.35
Biological oxygen demand w/w (days)	
Theoretical oxygen demand w/w	1.68

Vapour pressure equation constants (Log_{10}, mmHg)

Antoine equation	A	8.9545
	B	2692.2
	C	255.2
Cox chart	A	
	B	

Solvent properties

Solubility parameter		Kauri butanol value	
Dipole (D)		Evaporation time (ether = 1)	
Dielectric constant (20°C)		Evaporation time (BuAc = 1)	0.01
Polarity (water 100)	72.2		

Thermal information

Latent heat (cal/mol)	12844
Nett heat of combustion (kcal/gmol)	436
Specific heat (cal/mol/°C)	45
Critical pressure (MN/m^2)	6.1
Critical temperature (K)	624
Latent heat of fusion (cal/mol)	
Van der Waals' volume	3.28
Van der Waals' surface area	2.78
Molar volume	73.7

Solute	Azeotrope X% w/w	°C	Reference	Solute γ^{∞}	Reference	Partition coefficient	Reference
Hydrocarbons							
n-Pentane							
n-Hexane				120	1x/1/68		
n-Heptane				170	1x/1/68		
n-Octane				246	1x/1/68		
n-Nonane							
n-Decane							
2,2,4-TMP							
Cyclohexane				58.9	1x/1/68		
Benzene	None		6654	16.2	1x/1/68		
Toluene	2	110	6658	26.4	1x/1/68		
Ethylbenzene							
Xylenes	10	136	6665				
C_9 Aromatics							
Tetralin							
Alcohols							
Methanol							
Ethanol				1.1	2c/319		
n-Propanol				1.4	2c/491		
i-Propanol				1.5	2d/47		
n-Butanol				1.7	2d/137		
i-Butanol							
s-Butanol							
n-Amyl alc.				2.2	2d/139		
i-Amyl alc.				1.7	2d/138		
Cyclohexanol							
1-Octanol							
Ethanediol				0.9	2b/12		
DEG							
1,2-Propanediol	–	–	–	–	–	–	
Glycol ethers							
PGME							
EGME							
EEE							
EGBE							
Chlorinated							
MDC							
Chloroform							
Carbon tet.							
1,2-EDC							
1,1,1-TCA							
TCE							
Perk.							
MCB							

Solute	Azeotrope X% w/w	°C	Reference	Solute γ^∞	Reference	Partition coefficient	Reference
Ketones							
Acetone							
MEK							
MIBK							
Cyclohexanone							
NMP							
Acetophenone		183	6664				
Ethers							
Diethyl ether							
DIPE							
Dibutyl ether		136	6668				
MTBE							
1,4-Dioxane							
THF							
Esters							
Me acetate							
Et acetate				3.1	2d/135		
i-Propyl acetate							
n-Butyl acetate	None		6655a				
Cellosolve acetate	None		6656				
Miscellaneous							
DMF							
DMAc							
DMSO							
Sulfolane							
CS_2							
Acetic acid							
Aniline	43	180	6655				
Nitrobenzene							
Morpholine							
Pyridine							
2-Nitropropane							
Acetonitrile							
Furfuraldehyde							
Phenol							
Water	None		296	1.2	1x/3/1013		

Section 3

Glycol ethers

Propylene glycol methyl ether

Alternative names
1-Methoxy-2-propanol, PM, PGME

Reference codes

CAS number	107 98 2	Hazchem code	
UN number		EPA code	

Physical properties

Molecular weight	90	Cubic expansion coeff (per °C × 10^3)	1.03
Empirical formula	$C_4H_{10}O_2$	Surface tension (@20°C dyn/cm)	27.0
Boiling point (°C)	121	Absolute viscosity (@25°C cP)	1.9
Freezing point (°C)	–139	Refractive index (25°C)	1.407
Specific gravity (20/4)	0.924		

Fire hazards

Flash point (closed cup °C)	32	Lower explosive limit (ppm)	
Autoignition temperature (°C)	290	Upper explosive limit (ppm)	
Electrical conductivity	4.5E–7		

Health hazards

IDLH (ppm)		Vapour concentration @21°C ppm	11000
OES-TWA	100	Vapour density (relative to air)	3.1
OES-STEL	300	Vapour pressure @21°C mmHg	8.3
Odour threshold (ppm)		POCP	80

Aqueous effluent

Solubility in water (25°C %w/w)	Total
Solubility of water in (25°C %w/w)	Total
Log_{10} activated carbon partition	
Log_{10} partition in octanol/water (w/w)	
Biological oxygen demand w/w (days)	0.16 (5)
Theoretical oxygen demand w/w	1.95

Vapour pressure equation constants (Log_{10}, mmHg)

Antoine equation	A	
	B	
	C	
Cox chart	A	
	B	

Solvent properties

Solubility parameter	14.3	Kauri butanol value	
Dipole (D)	1.67	Evaporation time (ether = 1)	22
Dielectric constant (20°C)		Evaporation time (BuAc = 1)	0.66
Polarity (water 100)			

Thermal information

Latent heat (cal/mol)	9180
Nett heat of combustion (kcal/gmol)	682
Specific heat (cal/mol/°C)	49.5
Critical pressure (MN/m^2)	
Critical temperature (K)	552
Latent heat of fusion (cal/mol)	
Van der Waals' volume	
Van der Waals' surface area	
Molar volume	97.4

Solute	Azeotrope X% w/w	°C	Reference	Solute γ^{∞}	Reference	Partition coefficient	Reference
Hydrocarbons							
n-Pentane							
n-Hexane							
n-Heptane							
n-Octane							
n-Nonane							
n-Decane							
2,2,4-TMP							
Cyclohexane							
Benzene							
Toluene	30	107	8512				
Ethylbenzene							
Xylenes							
C_9 Aromatics							
Tetralin							
Alcohols							
Methanol							
Ethanol							
n-Propanol							
i-Propanol							
n-Butanol							
i-Butanol							
s-Butanol							
n-Amyl alc.							
i-Amyl alc.							
Cyclohexanol							
1-Octanol							
Ethanediol							
DEG							
1,2-Propanediol	None		6650a				
Glycol ethers							
PGME	–		–	–	–	–	–
EGME							
EEE							
EGBE							
Chlorinated							
MDC							
Chloroform							
Carbon tet.							
1,2-EDC							
1,1,1-TCA							
TCE							
Perk.							
MCB							

Glycol ethers

Solute	Azeotrope X% w/w	Azeotrope °C	Azeotrope Reference	Solute γ^∞	Reference	Partition coefficient	Reference
Ketones							
Acetone							
MEK							
MIBK							
Cyclohexanone							
NMP							
Acetophenone							
Ethers							
Diethyl ether							
DIPE							
Dibutyl ether							
MTBE							
1,4-Dioxane							
THF							
Esters							
Me acetate							
Et acetate							
i-Propyl acetate							
n-Butyl acetate							
Cellosolve acetate							
Miscellaneous							
DMF							
DMAc							
DMSO							
Sulfolane							
CS_2							
Acetic acid							
Aniline							
Nitrobenzene							
Morpholine							
Pyridine							
2-Nitropropane							
Acetonitrile							
Furfuraldehyde							
Phenol							
Water	65	97	384				

Ethylene glycol methyl ether

Alternative names
Methyl cellosolve, EGME, 2-methoxyethanol, ME, methyl glycol

Reference codes

CAS number	109 86 4	Hazchem code	2(S)
UN number	1188	EPA code	

Physical properties

Molecular weight	76	Cubic expansion coeff (per °C × 10^3)	0.92
Empirical formula	$C_3H_8O_2$	Surface tension (@20°C dyn/cm)	33.0
Boiling point (°C)	125	Absolute viscosity (@25°C cP)	1.6
Freezing point (°C)	–85	Refractive index (25°C)	1.400
Specific gravity (20/4)	0.966		

Fire hazards

Flash point (closed cup °C)	38	Lower explosive limit (ppm)	25000
Autoignition temperature (°C)	288	Upper explosive limit (ppm)	198000
Electrical conductivity	1.0E–6		

Health hazards

IDLH (ppm)	2000	Vapour concentration @21°C ppm	9300
OES-TWA	5	Vapour density (relative to air)	2.6
OES-STEL		Vapour pressure @21°C mmHg	7
Odour threshold (ppm)	90	POCP	

Aqueous effluent

Solubility in water (25°C %w/w)	Total
Solubility of water in (25°C %w/w)	Total
Log_{10} activated carbon partition	1.50
Log_{10} partition in octanol/water (w/w)	–0.77
Biological oxygen demand w/w (days)	0.50 (5)
Theoretical oxygen demand w/w	1.68

Vapour pressure equation constants (Log_{10}, mmHg)

Antoine equation	A	7.8498
	B	1793.982
	C	236.877
Cox chart	A	
	B	

Solvent properties

Solubility parameter	10.8	Kauri butanol value	
Dipole (D)	2.0	Evaporation time (ether = 1)	34
Dielectric constant (20°C)	16.9	Evaporation time (BuAc = 1)	0.47
Polarity (water 100)	66.7		

Thermal information

Latent heat (cal/mol)	9424
Nett heat of combustion (kcal/gmol)	399
Specific heat (cal/mol/°C)	43
Critical pressure (MN/m^2)	5.1
Critical temperature (K)	565
Latent heat of fusion (cal/mol)	
Van der Waals' volume	
Van der Waals' surface area	
Molar volume	78.7

Solute	Azeotrope X% w/w	Azeotrope °C	Reference	Solute γ^{∞}	Reference	Partition coefficient	Reference
Hydrocarbons							
n-Pentane							
n-Hexane				13.1	1x/3/1012		
n-Heptane	23	93	6592	4.4	1x/1/67		
n-Octane	48	110	6614	28.4	1x/3/1012		
n-Nonane							
n-Decane	92	123	6628				
2,2,4-TMP							
Cyclohexane	15	78	6572	5.7	2b/128		
Benzene	None		6567	2.3	2b/127		
Toluene	25	106	6586	3.6	1x/3/1012		
Ethylbenzene	51	117	6596	2.9	2b/132		
Xylenes	55	120	6598	3.1	2b/134		
C_9 Aromatics				3.8	2d/131		
Tetralin							
Alcohols							
Methanol	None		1979				
Ethanol	None		3982	1.1	1x/3/1012		
n-Propanol							
i-Propanol				1.7	2c/490		
n-Butanol	None		6546				
i-Butanol	None		6548				
s-Butanol	None		6547				
n-Amyl aAlc.	None		6560				
i-Amyl alc.	None		6562				
Cyclohexanol							
1-Octanol							
Ethanediol							
DEG							
1,2-Propanediol							
Glycol ethers							
PGME							
EGME	–		–	–	–	–	–
EEE							
EGBE							
Chlorinated							
MDC							
Chloroform							
Carbon tet.				3.1	2f/99		
1,2-EDC							
1,1,1-TCA							
TCE							
Perk.	24	109	2178				
MCB	47	119	6566	2.4	2d/120		

Solute	Azeotrope X w/w	°C	Reference	Solute γ^{∞}	Reference	Partition coefficient	Reference
Ketones							
Acetone				1.6	2d/113		
MEK				2.0	2b/122		
MIBK	25	114	6575				
Cyclohexanone							
NMP							
Acetophenone							
Ethers							
Diethyl ether							
DIPE							
Dibutyl ether	68	122	6615				
MTBE							
1,4-Dioxane	None		6541	1.5	1x/3/1012		
THF							
Esters							
Me acetate							
Et acetate				1.7	2b/126		
i-Propyl acetate							
n-Butyl acetate	48	119	6576	2.4	2d/122		
Cellosolve acetate	None		6583				
Miscellaneous							
DMF							
DMAc							
DMSO							
Sulfolane				2.2	2f/103		
CS_2							
Acetic acid							
Aniline							
Nitrobenzene							
Morpholine							
Pyridine	None		6550				
2-Nitropropane							
Acetonitrile				1.7	2d/109		
Furfuraldehyde	None		6549				
Phenol	None		6568				
Water	19	99	294				

Ethylene glycol ethyl ether

Alternative names

Cellosolve, EGEE, 2-ethoxyethanol, EE, ethyl glycol

Reference codes

CAS number	110 80 5	Hazchem code	2S
UN number	1711	EPA code	

Physical properties

Molecular weight	90	Cubic expansion coeff (per °C × 10^3)	0.97
Empirical formula	$C_4H_{10}O_2$	Surface tension (@20°C dyn/cm)	28.2
Boiling point (°C)	135	Absolute viscosity (@25°C cP)	2.5
Freezing point (°C)	–70	Refractive index (25°C)	1.405
Specific gravity (20/4)	0.931		

Fire hazards

Flash point (closed cup °C)	43	Lower explosive limit (ppm)	18000
Autoignition temperature (°C)	235	Upper explosive limit (ppm)	140000
Electrical conductivity	9.3E–8		

Health hazards

IDLH (ppm)	6000	Vapour concentration @21°C ppm	5300
OES-TWA	5	Vapour density (relative to air)	3.1
OES-STEL		Vapour pressure @21°C mmHg	4
Odour threshold (ppm)	50	POCP	75

Aqueous effluent

Solubility in water (25°C %w/w)	Total
Solubility of water in(25°C %w/w)	Total
Log_{10} activated carbon partition	1.95
Log_{10} partition in octanol/water (w/w)	–0.54
Biological oxygen demand w/w (days)	0.67 (5)
Theoretical oxygen demand w/w	1.86

Vapour pressure equation constants (Log_{10}, mmHg)

Antoine equation	A	7.81910
	B	1801.90
	C	230
Cox chart	A	
	B	

Solvent properties

Solubility parameter	10.0	Kauri butanol value	
Dipole (D)	1.69	Evaporation time (ether = 1)	43
Dielectric constant (20°C)	5.3	Evaporation time (BuAc = 1)	0.32
Polarity (water 100)	62.7		

Thermal information

Latent heat (cal/mol)	9540
Nett heat of combustion (kcal/gmol)	503
Specific heat (cal/mol/°C)	52
Critical pressure (MN/m^2)	
Critical temperature (K)	
Latent heat of fusion (cal/mol)	
Van der Waals' volume	3.70
Van der Waals' surface area	3.29
Molar volume	97.41

Solute	Azeotrope X% w/w	°C	Reference	Solute γ^∞	Reference	Partition coefficient	Reference
Hydrocarbons							
n-Pentane							
n-Hexane	5	66	8442	6.0	2b/295		
n-Heptane	14	97	8461				
n-Octane	38	116	8478	5.5	2b/302		
n-Nonane	50	128	8493				
n-Decane							
2,2,4-TMP							
Cyclohexane							
Benzene	None		8425				
Toluene	11	109	8450	2.0	2f/337		
Ethylbenzene	43	126	8463	2.4	2d/299		
Xylenes	50	128	8465	2.9	2f/338		
C_9 Aromatics							
Tetralin							
Alcohols							
Methanol							
Ethanol	None		4032				
n-Propanol							
i-Propanol							
n-Butanol	None		8105				
i-Butanol	None		8306				
s-Butanol							
n-Amyl alc.	None		8419				
i-Amyl alc.	None		8420				
Cyclohexanol							
1-Octanol							
Ethanediol							
DEG							
1,2-Propanediol							
Glycol ethers							
PGME							
EGME							
EEE	–		–	–	–	–	–
EGBE							
Chlorinated							
MDC							
Chloroform							
Carbon tet.							
1,2-EDC							
1,1,1-TCA							
TCE							
Perk.	16	116	2190		2d/396		
MCB	32	127	8423				

Glycol ethers

Solute	Azeotrope X% w/w	Azeotrope °C	Reference	Solute γ^{∞}	Reference	Partition coefficient	Reference
Ketones							
Acetone				1.0	2f/332		
MEK				1.6	2f/334		
MIBK	None		8433				
Cyclohexanone							
NMP							
Acetophenone							
Ethers							
Diethyl ether							
DIPE							
Dibutyl ether	50	127	8479				
MTBE							
1,4-Dioxane							
THF							
Esters							
Me acetate							
Et acetate	None		7571	1.5	2f/335		
i-Propyl acetate							
n-Butyl acetate	13	126	8434	1.6	2b/294		
Cellosolve acetate	None		8400				
Miscellaneous							
DMF							
DMAc							
DMSO							
Sulfolane							
CS_2							
Acetic acid							
Aniline							
Nitrobenzene							
Morpholine							
Pyridine	None		8407				
2-Nitropropane							
Acetonitrile	15	119	6279				
Furfuraldehyde	None		8406				
Phenol	None		8426				
Water	13	98	382	1.9	1/450		

Ethylene glycol monobutyl ether

Alternative names
Butyl glycol, butyl cellosolve, EB, 2-butoxyethanol, EGBE

Reference codes

CAS number	111 76 2	Hazchem code	2R
UN number	2369	EPA code	

Physical properties

Molecular weight	118	Cubic expansion coeff (per °C × 10^3)	0.92
Empirical formula	$C_6H_{14}O_2$	Surface tension (@20°C dyn/cm)	27.4
Boiling point (°C)	171	Absolute viscosity (@25°C cP)	6.4
Freezing point (°C)	–75	Refractive index (25°C)	1.417
Specific gravity (20/4)	0.902		

Fire hazards

Flash point (closed cup °C)	68	Lower explosive limit (ppm)	11000
Autoignition temperature (°C)	214	Upper explosive limit (ppm)	106000
Electrical conductivity	4.3E–7		

Health hazards

IDLH (ppm)	700	Vapour concentration @21°C ppm	922
OES-TWA	25	Vapour density (relative to air)	4.07
OES-STEL		Vapour pressure @21°C mmHg	0.7
Odour threshold (ppm)	0.5	POCP	75

Aqueous effluent

Solubility in water (25°C %w/w)	Total } LCST 58°C
Solubility of water in (25°C %w/w)	Total } LCST 58°C
Log_{10} activated carbon partition	2.40
Log_{10} partition in octanol/water (w/w)	+0.83
Biological oxygen demand w/w (days)	0.60 (5)
Theoretical oxygen demand w/w	2.3

Vapour pressure equation constants (Log_{10}, mmHg)

Antoine equation	A	7.8448
	B	1988.90
	C	230.00
Cox chart	A	
	B	

Solvent properties

Solubility parameter	8.9	Kauri butanol value	
Dipole (D)	1.80	Evaporation time (ether = 1)	119
Dielectric constant (20°C)	5.3	Evaporation time (BuAc = 1)	0.06
Polarity (water 100)	60.2		

Thermal information

Latent heat (cal/mol)	10266
Nett heat of combustion (kcal/gmol)	778
Specific heat (cal/mol/°C)	55
Critical pressure (MN/m^2)	3.2
Critical temperature (K)	641
Latent heat of fusion (cal/mol)	
Van der Waals' volume	5.05
Van der Waals' surface area	4.37
Molar volume	131.84

Solute	Azeotrope X% w/w	°C	Reference	Solute γ^∞	Reference	Partition coefficient	Reference
Hydrocarbons							
n-Pentane							
n-Hexane							
n-Heptane							
n-Octane	None			3.3	2b/432		
n-Nonane							
n-Decane							
2,2,4-TMP							
Cyclohexane							
Benzene							
Toluene				1.3	2f/440		
Ethylbenzene	96	140					
Xylenes	None		12235				
C_9 Aromatics	30	160	12252				
Tetralin							
Alcohols							
Methanol							
Ethanol							
n-Propanol							
i-Propanol							
n-Butanol	None		8166	1.1	2f/89		
i-Butanol							
s-Butanol							
n-Amyl alc.							
i-Amyl alc.							
Cyclohexanol	None		11712				
1-Octanol							
Ethanediol							
DEG							
1,2-Propanediol							
Glycol ethers							
PGME							
EGME	–		–	–	–	–	–
EEE							
EGBE							
Chlorinated							
MDC							
Chloroform							
Carbon tet.							
1,2-EDC							
1,1,1-TCA							
TCE							
Perk.	None		2218				
MCB	None		10521				

Glycol ethers

Solute	Azeotrope X% w/w	°C	Reference	Solute γ^∞	Reference	Partition coefficient	Reference
Ketones							
Acetone							
MEK				1.4	2b/430		
MIBK							
Cyclohexanone							
NMP							
Acetophenone							
Ethers							
Diethyl ether							
DIPE							
Dibutyl ether	None		12247				
MTBE							
1,4-Dioxane							
THF							
Esters							
Me acetate							
Et acetate	None		7590				
i-Propyl acetate							
n-Butyl acetate	None		11823				
Cellosolve acetate	None		11989				
Miscellaneous							
DMF							
DMAc							
DMSO							
Sulfolane							
CS_2							
Acetic acid							
Aniline	None		11153				
Nitrobenzene	None		10710				
Morpholine							
Pyridine							
2-Nitropropane							
Acetonitrile							
Furfuraldehyde	12	161	8769				
Phenol	37	186	10904				
Water	21	99	584	0.73	1/526		

Section 4

Chlorinated solvents

Methylene chloride

Alternative names
Dichloromethane, MDC, methylene dichloride, **not** methyl chloride

Reference codes

CAS number	75 09 2	Hazchem code	2Z
UN number	1593	EPA code	U080

Physical properties

Molecular weight	85	Cubic expansion coeff (per °C × 10^3)	1.37
Empirical formula	$C_1H_2Cl_2$	Surface tension (@20°C dyn/cm)	28.1
Boiling point (°C)	40	Absolute viscosity (@25°C cP)	0.44
Freezing point (°C)	–95	Refractive index (25°C)	1.4211
Specific gravity (20/4)	1.326		

Fire hazards

Flash point (closed cup °C)	None	Lower explosive limit (ppm)	130000
Autoignition temperature (°C)	605	Upper explosive limit (ppm)	220000
Electrical conductivity	4.3E–11		

Health hazards

IDLH (ppm)	5000	Vapour concentration @21°C ppm	495000
OES-TWA	50	Vapour density (relative to air)	2.95
OES-STEL	300	Vapour pressure @21°C mmHg	376
Odour threshold (ppm)	250	POCP	0.9

Aqueous effluent

Solubility in water (25°C %w/w)	1.30	
Solubility of water in (25°C %w/w)	0.20	
Log_{10} activated carbon partition	2.9	
Log_{10} partition in octanol/water (w/w)	+1.25	
Biological oxygen demand w/w (days)		0 (5)
Theoretical oxygen demand w/w	0.56	

Vapour pressure equation constants (Log_{10}, mmHg)

Antoine equation	A	7.0803
	B	1138.91
	C	231.45
Cox chart	A	6.91821
	B	1090.1

Solvent properties

Solubility parameter	9.7	Kauri butanol value	136
Dipole (D)	1.8	Evaporation time (ether = 1)	1.8
Dielectric constant (20°C)	9.1	Evaporation time (BuAc = 1)	25.0
Polarity (water 100)	30.9		

Thermal information

Latent heat (cal/mol)	6715
Nett heat of combustion (kcal/gmol)	122
Specific heat (cal/mol/°C)	24
Critical pressure (MN/m^2)	6.08
Critical temperature (K)	510
Latent heat of fusion (cal/mol)	1436
Van der Waals' volume	2.26
Van der Waals' surface area	1.99
Molar volume	64.50

Solute	Azeotrope X% w/w	°C	Reference	Solute γ^∞	Reference	Partition coefficient	Reference
Hydrocarbons							
n-Pentane	49	36	1571	2.9	6a/100		
n-Hexane	None		1575				
n-Heptane							
n-Octane				4.1	1x/3/923		
n-Nonane							
n-Decane							
2,2,4-TMP							
Cyclohexane							
Benzene							
Toluene				1.0	1x/3/923		
Ethylbenzene							
Xylenes							
C_9 Aromatics							
Tetralin							
Alcohols							
Methanol	93	38	1544	7.9	2e/24	9.28	V4/118
Ethanol	95	40	1551	43.7	2c/283		
n-Propanol				4.1	2e/416		
i-Propanol	None		1561	4.1	2f/36		
n-Butanol							
i-Butanol							
s-Butanol							
n-Amyl alc.							
i-Amyl alc.							
Cyclohexanol							
1-Octanol							
Ethanediol							
DEG							
1,2-Propanediol							
Glycol ethers							
PGME							
EGME							
EEE							
EGBE							
Chlorinated							
MDC	–		–	–	–	–	–
Chloroform	None		1426	0.7	8/202		
Carbon tet.				0.9	8/62		
1,2-EDC				1.0	8/263		
1,1,1-TCA				1.2	1x/3/923		
TCE				1.5	1x/3/923		
Perk.				0.9	8/256		
MCB							

Solute	Azeotrope			Solute γ^{∞}	Reference	Partition coefficient	Reference
	X% w/w	°C	Reference				
Ketones							
Acetone	None		1553	1.1	3b/27		
MEK	None		1564	0.4	3+4/261		
MIBK							
Cyclohexanone							
NMP							
Acetophenone							
Ethers							
Diethyl ether	70	41	1566	0.7	3+4/492		
DIPE							
Dibutyl ether							
MTBE							
1,4-Dioxane				0.4	1x/3/923		
THF					0.09		
Esters							
Me acetate	None		1557	0.5	5/347		
Et acetate				0.4	5/449		
i-Propyl acetate							
n-Butyl acetate							
Cellosolve acetate							
Miscellaneous							
DMF				0.8	8/265	0.39	V4/120
DMAc							
DMSO				0.45	8/264		
Sulfolane				0.8	8/266		
CS_2	65	36	1170				
Acetic acid				4.2	5/64	3.24	V2/36
Aniline							
Nitrobenzene							
Morpholine							
Pyridine				0.6	8/267		
2-Nitropropane							
Acetonitrile	None		1546	1.2	8/258		
Furfuraldehyde				1.2	3a/115		
Phenol							
Water	99	38	208	1324.0	1/1		

Chloroform

Alternative names
Trichloromethane

Reference codes

CAS number	67 66 3	Hazchem code	2Z
UN number	1888	EPA code	U044

Physical properties

Molecular weight	119	Cubic expansion coeff (per °C × 10^3)	1.29
Empirical formula	$C_1H_1Cl_3$	Surface tension (@20°C dyn/cm)	27.16
Boiling point (°C)	61	Absolute viscosity (@25°C cP)	0.57
Freezing point (°C)	–23	Refractive index (25°C)	1.444
Specific gravity (20/4)	1.480		

Fire hazards

Flash point (closed cup °C)	None	Lower explosive limit (ppm)	None
Autoignition temperature (°C)	None	Upper explosive limit (ppm)	None
Electrical conductivity	<1.0E–10		

Health hazards

IDLH (ppm)	1000	Vapour concentration @21°C ppm	286000
OES-TWA	2	Vapour density (relative to air)	4.13
OES-STEL		Vapour pressure @21°C mmHg	169
Odour threshold (ppm)	300	POCP	1.0

Aqueous effluent

Solubility in water (25°C %w/w)	0.82
Solubility of water in (25°C %w/w)	0.072
Log_{10} activated carbon partition	3.6
Log_{10} partition in octanol/water (w/w)	+1.97
Biological oxygen demand w/w (days)	0.02 (5)
Theoretical oxygen demand w/w	1.35

Vapour pressure equation constants (Log_{10}, mmHg)

Antoine equation	A	6.95465
	B	1170.966
	C	226.232
Cox chart	A	6.97909
	B	1192.6

Solvent properties

Solubility parameter	9.3	Kauri butanol value	
Dipole (D)	1.1	Evaporation time (ether = 1)	1.9
Dielectric constant (20°C)	4.8	Evaporation time (BuAc = 1)	
Polarity (water 100)	25.9		

Thermal information

Latent heat (cal/mol)	70.21
Nett heat of combustion (kcal/gmol)	91
Specific heat (cal/mol/°C)	27
Critical pressure (MN/m^2)	2.38
Critical temperature (K)	536
Latent heat of fusion (cal/mol)	2097
Van der Waals' volume	2.87
Van der Waals' surface area	2.41
Molar volume	80.41

Solute	Azeotrope X% w/w	Azeotrope °C	Reference	Solute γ^∞	Reference	Partition coefficient	Reference
Hydrocarbons							
n-Pentane	None		1482	2.1	1x/3/992		
n-Hexane	83	60	1495	1.9	6a/426		
n-Heptane	None		1500	1.4	6b/77		
n-Octane				2.1	1x/3/922		
n-Nonane							
n-Decane							
2,2,4-TMP							
Cyclohexane	None		1490				
Benzene	None		1486	0.86	1x/1/4		
Toluene	None		1498	0.75	7/352		
Ethylbenzene							
Xylenes							
C_9 Aromatics							
Tetralin							
Alcohols							
Methanol	87	53	1430	7.4	2a/23	3.5	P160
Ethanol	93	59	1442	4.3	2a/285	1.07	P373
n-Propanol	None		1454			0.24	P640
i-Propanol	96	61	1453	6.6	2d/40	0.34	P651
n-Butanol				2.7	2b/136	0.05	P952
i-Butanol	None		1475			0.07	P964
s-Butanol	None		1472			0.08	P976
n-Amyl alc.							
i-Amyl alc.							
Cyclohexanol							
1-Octanol							
Ethanediol							
DEG							
1,2-Propanediol							
Glycol ethers							
PGME							
EGME							
EEE							
EGBE							
Chlorinated							
MDC	None		1426	0.8	8/202		
Chloroform	–		–	–	–	–	–
Carbon tet.	None		1086	1.1	8/56		
1,2-EDC	None		1435	1.1	1x/3/921		
1,1,1-TCA				1.0	1x/3/921		
TCE				1.1	1x/3/921		
Perk.	None		1431	1.2	8/215		
MCB	None		1484	0.8	8/244		

Solute	Azeotrope X% w/w	°C	Reference	Solute γ^{∞}	Reference	Partition coefficient	Reference
Ketones							
Acetone	78	64	1443	0.5	3+4/90	0.03	P493
MEK	17	80	1460	0.4	3+4/260		
MIBK	None		1492	0.3	3+4/343		
Cyclohexanone							
NMP				0.06	3b/426		
Acetophenone							
Ethers							
Diethyl ether	None		1474	0.4	3+4/486		
DIPE	36	71	1496	0.5	3+4/537		
Dibutyl ether	None		1501a	0.4	3+4/591		
MTBE							
1,4-Dioxane	None		1465	0.3	3+4/441		
THF	66	73	1464	0.25	1x/1/4	0.03	V4/115
Esters							
Me acetate	77	65	1448	0.43	5/341		
Et acetate	28	78	1466	0.23	5/443		
i-Propyl acetate							
n-Butyl acetate	None		1493	0.41	5/574		
Cellosolve acetate							
Miscellaneous							
DMF						0.32	V4/113
DMAc							
DMSO				0.18	8/229		
Sulfolane							
CS_2	None		1169	1.3	8/213		
Acetic acid	None		1437	4.2	5/62	1.04	V2/18
Aniline						0.007	P1714
Nitrobenzene	None		1485				
Morpholine							
Pyridine	None		1480a	0.35	8/240	0.006	P1102
2-Nitropropane							
Acetonitrile	None		1433	1.2	8/217		
Furfuraldehyde	None		1480	10.7	3+4/36		
Phenol						0.07	P1627
Water	97	56	207				

Carbon tetrachloride

Alternative names
Carbon tet., CTET, tetrachloromethane

Reference codes

CAS number	56 23 5	Hazchem code	2Z
UN number	1846	EPA code	U211

Physical properties

Molecular weight	154	Cubic expansion coeff (per °C × 10^3)	1.27
Empirical formula	C_1Cl_4	Surface tension (@20°C dyn/cm)	27
Boiling point (°C)	76	Absolute viscosity (@25°C cP)	0.97
Freezing point (°C)	–23	Refractive index (25°C)	1.459
Specific gravity (20/4)	1.58		

Fire hazards

Flash point (closed cup °C)	None	Lower explosive limit (ppm)	None
Autoignition temperature (°C)	None	Upper explosive limit (ppm)	None
Electrical conductivity	4.0E–18		

Health hazards

IDLH (ppm)	300	Vapour concentration @21°C ppm	150900
OES-TWA	2	Vapour density (relative to air)	5.34
OES-STEL		Vapour pressure @21°C mmHg	99.6
Odour threshold (ppm)	96	POCP	

Aqueous effluent

Solubility in water (25°C %w/w)	0.077
Solubility of water in (25°C %w/w)	0.008
Log_{10} activated carbon partition	4.3
Log_{10} partition in octanol/water (w/w)	+2.64
Biological oxygen demand w/w (days)	0 (5)
Theoretical oxygen demand w/w	0.21

Vapour pressure equation constants (Log_{10}, mmHg)

Antoine Equation	A	6.84083
	B	1210.595
	C	229.664
Cox chart	A	7.02433
	B	1267.9

Solvent properties

Solubility parameter	8.6	Kauri butanol value	
Dipole (D)	0	Evaporation time (ether = 1)	1.8
Dielectric constant (20°C)	2.24	Evaporation Time (BuAc = 1)	
Polarity (water 100)	5.2		

Thermal information

Latent heat (cal/mol)	7238
Nett heat of combustion (kcal/gmol)	62
Specific heat (cal/mol/°C)	32
Critical pressure (MN/m^2)	4.56
Critical temperature (K)	556
Latent heat of fusion (cal/mol)	784
Van der Waals' volume	3.39
Van der Waals' surface area	2.91
Molar volume	97.5

Solute	Azeotrope X% w/w	Azeotrope °C	Azeotrope Reference	Solute γ^∞	Reference	Partition coefficient	Reference
Hydrocarbons							
n-Pentane							
n-Hexane	None		1162	1.3	6a/403		
n-Heptane	None		1167	1.3	6b/67		
n-Octane	None		1168a	1.0	6b/234		
n-Nonane							
n-Decane							
2,2,4-TMP	None		1168b	1.4	6b/285		
Cyclohexane	None		1159	1.1	6a/142		
Benzene	None		1154	1.1	7/7		
Toluene	None		1166	1.0	7/332		
Ethylbenzene	None		1167a	0.9	7/464		
Xylenes	None		1167c	0.9	7/480		
C_9 Aromatics							
Tetralin							
Alcohols							
Methanol	79	56	1090	10.3	2a/1		
Ethanol	84	65	1105	20.5	2a/276	4.76	P382
n-Propanol	92	73	1116	14.5	2a/509		
i-Propanol	82	67	1115	6.7	2b/36	2.45	V2/10
n-Butanol	98	77	1133	9.5	2f/109	0.32	P958
i-Butanol	95	76	1137				
s-Butanol	92	75	1134	5.4	2f/217		
n-Amyl alc.	None		1147a			0.05	P1269
i-Amyl alc.	None		1149				
Cyclohexanol							
1-Octanol							
Ethanediol							
DEG							
1,2-Propanediol							
Glycol ethers							
PGME							
EGME				4.9	2f/99		
EEE							
EGBE							
Chlorinated							
MDC				1.1	8/63		
Chloroform	None		1086	1.0	8/55		
Carbon tet.	–		–	–	–	–	–
1,2-EDC	80	75	1098	1.5	8/96		
1,1,1-TCA	Azeo			1.2	8/82		
TCE	None		1093	0.7	8/80		
Perk.	None		1091	1.0	8/78		
MCB	None		1152	1.4	8/164		

Chlorinated solvents

Solute	Azeotrope X% w/w	°C	Reference	Solute γ^∞	Reference	Partition coefficient	Reference
Ketones							
Acetone	12	56	1108	2.3	3+4/80	0.26	P502
MEK	71	74	1121	1.8	3+4/259		
MIBK							
Cyclohexanone				0.9	3b/495		
NMP							
Acetophenone							
Ethers							
Diethyl ether	None		1136				
DIPE	None		1163	1.1	3+4/529		
Dibutyl ether							
MTBE							
1,4-Dioxane	None		1124	1.6	3+4/440		
THF				0.9	3+4/429	0.06	V4/107
Esters							
Me acetate	None		1110	1.6	5/339	0.05	P519
Et acetate	57	75	1125	1.4	5/436		
i-Propyl acetate	None		1145				
n-Butyl acetate	None		1161	0.8	5/573		
Cellosolve acetate							
Miscellaneous							
DMF				5.4	8/117	5.9	V4/106
DMAc							
DMSO				18.5	8/107	3.8	P387
Sulfolane							
CS_2	None		1085	1.2	8/76		
Acetic acid	98	76	1099	12.2	5/59	6.0	V2/12
Aniline	None			6.0	8/174	0.07	P1720
Nitrobenzene	None		1153	0.7	8/168		
Morpholine							
Pyridine				2.0	8/141		
2-Nitropropane	None		1114	3.2	8/120		
Acetonitrile	83	65	1095	9.6	8/86		
Furfuraldehyde	None		1140	3.8	3+4/35	0.04	
Phenol	None		1155	3.4	2b/355	0.31	P1652
Water	96	66	205				

1,2-Dichloroethane

Alternative names
Ethylene dichloride, EDC

Reference codes

CAS number	107 06 2	Hazchem code	2YE
UN number	1184	EPA code	U077

Physical properties

Molecular weight	99	Cubic expansion coeff (per °C × 10^3)	1.16
Empirical formula	$C_2H_4Cl_2$	Surface tension (@20°C dyn/cm)	32.2
Boiling point (°C)	83.5	Absolute viscosity (@25°C cP)	0.9
Freezing point (°C)	–36	Refractive index (25°C)	1.444
Specific gravity (20/4)	1.253		

Fire hazards

Flash point (closed cup °C)	13	Lower explosive limit (ppm)	62000
Autoignition temperature (°C)	413	Upper explosive limit (ppm)	169000
Electrical conductivity	4E–11		

Health hazards

IDLH (ppm)	1000	Vapour concentration @21°C ppm	94000
OES-TWA	5	Vapour density (relative to air)	3.4
OES-STEL		Vapour pressure @21°C mmHg	71
Odour threshold (ppm)	400	POCP	

Aqueous effluent

Solubility in water (25°C %w/w)	0.81
Solubility of water in (25°C %w/w)	0.15
Log_{10} activated carbon partition	3.8
Log_{10} partition in octanol/water (w/w)	3.6
Biological oxygen demand w/w (days)	0.002 (5)
Theoretical oxygen demand w/w	0.97

Vapour pressure equation constants (Log_{10}, mmHg)

Antoine equation	A	7.02530
	B	1271.254
	C	222.927
Cox chart	A	7.04532
	B	1303.5

Solvent properties

Solubility parameter	9.8	Kauri butanol value	
Dipole (D)	1.8	Evaporation time (ether = 1)	2.7
Dielectric constant (20°C)	10.45	Evaporation time (BuAc = 1)	
Polarity (water 100)	32.7		

Thermal information

Latent heat (cal/mol)	7623
Nett heat of combustion (kcal/gmol)	269
Specific heat (cal/mol/°C)	31
Critical pressure (MN/m^2)	5.38
Critical temperature (K)	563
Latent heat of fusion (cal/mol)	2091
Van der Waals' volume	2.93
Van der Waals' surface area	2.53
Molar volume	78.87

Solute	Azeotrope X% w/w	°C	Reference	Solute γ^{∞}	Reference	Partition coefficient	Reference
Hydrocarbons							
n-Pentane				4.8	1x/1/18		
n-Hexane	None		3003	3.6	1x/1/19		
n-Heptane	76	81	3009	3.4	6c/444		
n-Octane							
n-Nonane							
n-Decane							
2,2,4-TMP				3.4	1x/1/19		
Cyclohexane	50	74	3001	3.1	6a/159		
Benzene	20	80	2999	1.1	7/142		
Toluene	None		3006	1.1	7/380		
Ethylbenzene				1.0	7/466		
Xylenes				1.3	7/490		
C_9 Aromatics							
Tetralin							
Alcohols							
Methanol	68	61	1930	11.2	2e/44	7.9	V2/83
Ethanol	63	71	2964	4.9	2a/299		
n-Propanol	None		2971	5.1	2a/520		
i-Propanol	57	73	2970	4.9	1x/1/18	1.2	V2/207
n-Butanol	None		2984	3.2	2b/137		
i-Butanol	94	83	2988	4.2	2b/272		
s-Butanol	88	82	2985	2.7	2f/220		
n-Amyl alc.				3.6	2f/375		
i-Amyl alc.	None		2995				
Cyclohexanol							
1-Octanol							
Ethanediol							
DEG							
1,2-Propanediol							
Glycol ethers							
PGME							
EGME							
EEE							
EGBE							
Chlorinated							
MDC				1.0	8/263		
Chloroform	None		1435	1.0	1x/3/957		
Carbon tet.	20	75	1098	1.8	1x/3/957		
1,2-EDC	–		–	–	–	–	–
1,1,1-TCA				1.6	8/363		
TCE	67	82	2281	1.4	8/351		
Perk.				1.7	8/340		
MCB							

Solute	Azeotrope X% w/w	Azeotrope °C	Azeotrope Reference	Solute γ^{∞}	Reference	Partition coefficient	Reference
Ketones							
Acetone	None		2966	0.8	3+4/144		
MEK	Azeo		2977	0.8	3b/271		
MIBK				0.8	3b/519		
Cyclohexanone							
NMP							
Acetophenone						<0.01	P2746
Ethers							
Diethyl ether	None		2987				
DIPE	None		3004				
Dibutyl ether							
MTBE							
1,4-Dioxane	None		2979	0.9	3+4/447	0.98	V4/180
THF				0.7	1x/3/957		
Esters							
Me acetate							
Et acetate	None		2980	0.8	1x/1/18		
i-Propyl acetate	None		2992				
n-Butyl acetate							
Cellosolve acetate							
Miscellaneous							
DMF							
DMAc							
DMSO							
Sulfolane							
CS_2				2.6	1x/1/17		
Acetic acid	None		2961	4.4	5/74	1.61	V2/202
Aniline						<0.01	P1721
Nitrobenzene							
Morpholine							
Pyridine				0.9	1x/1/18		
2-Nitropropane							
Acetonitrile	51	79	2757	1.4	8/364		
Furfuraldehyde				1.1	3a/119	0.14	V4/181
Phenol							
Water	91	72	227				

1,1,1-Trichloroethane

Alternative names
Methyl chloroform, TCA, chlorothene, M.C., **not** 1,1,2-trichloroethane

Reference codes

CAS number	71 55 6	Hazchem code	2Z
UN number	2831	EPA code	U226

Physical properties

Molecular weight	133	Cubic expansion coeff (per °C × 10^3)	1.3
Empirical formula	$C_2H_3Cl_3$	Surface tension (@20°C dyn/cm)	30
Boiling point (°C)	74	Absolute viscosity (@25°C cP)	0.65
Freezing point (°C)	–30	Refractive index (25°C)	1.438
Specific gravity (20/4)	1.338		

Fire hazards

Flash point (closed cup °C)	None	Lower explosive limit (ppm)	65000
Autoignition temperature (°C)	537	Upper explosive limit (ppm)	155000
Electrical conductivity	7.3E–9		

Health hazards

IDLH (ppm)		Vapour concentration @21°C ppm	154700
OES-TWA	350	Vapour density (relative to air)	4.62
OES-STEL	450	Vapour pressure @21°C mmHg	101.8
Odour threshold (ppm)	300	POCP	0.1

Aqueous effluent

Solubility in water (25°C %w/w)	0.13
Solubility of water in (25°C %w/w)	0.03
Log_{10} activated carbon partition	4.3
Log_{10} partition in octanol/water (w/w)	4.0
Biological oxygen demand w/w (days)	
Theoretical oxygen demand w/w	0.48

Vapour pressure equation constants (Log_{10}, mmHg)

Antoine equation	A	6.90633
	B	1211.31
	C	226.816
Cox chart	A	7.01846
	B	1257.7

Solvent properties

Solubility parameter	7.7	Kauri butanol value	124
Dipole (D)	1.7	Evaporation time (ether = 1)	2.6
Dielectric constant (20°C)	7.25	Evaporation time (BuAc = 1)	6.0
Polarity (water 100)	17.0		

Thermal information

Latent heat (cal/mol)	7780
Nett heat of combustion (kcal/gmol)	233
Specific heat (cal/mol/°C)	32
Critical pressure (MN/m^2)	4.4
Critical temperature (K)	550
Latent heat of fusion (cal/mol)	651.7
Van der Waals' volume	3.54
Van der Waals' surface area	3.03
Molar volume	100.4

Solute	Azeotrope X% w/w	°C	Reference	Solute γ^{∞}	Reference	Partition coefficient	Reference
Hydrocarbons							
n-Pentane							
n-Hexane	29	67	2730b	1.3	6a/473		
n-Heptane							
n-Octane							
n-Nonane							
n-Decane							
2,2,4-TMP							
Cyclohexane							
Benzene				1.0	7/121		
Toluene							
Ethylbenzene							
Xylenes							
C_9 Aromatics							
Tetralin							
Alcohols							
Methanol	78	56	1923				
Ethanol							
n-Propanol							
i-Propanol							
n-Butanol				1.6	2f/123	0.11	V4/158
i-Butanol							
s-Butanol							
n-Amyl alc.							
i-Amyl alc.				24.8	2f/373		
Cyclohexanol							
1-Octanol							
Ethanediol							
DEG							
1,2-Propanediol							
Glycol ethers							
PGME							
EGME							
EEE							
EGBE							
Chlorinated							
MDC							
Chloroform				1.0	1x/3/948		
Carbon tet.	Azeo			1.1	8/82		
1,2-EDC	None		2729	1.2	1x/3/948		
1,1,1-TCA	–		–	–	–	–	–
TCE				1.0	1x/3/948		
Perk.							
MCB							

Chlorinated solvents

Solute	Azeotrope			Solute γ^{∞}	Reference	Partition coefficient	Reference
	X% w/w	°C	Reference				
Ketones							
Acetone							
MEK							
MIBK							
Cyclohexanone							
NMP							
Acetophenone							
Ethers							
Diethyl ether							
DIPE							
Dibutyl ether							
MTBE							
1,4-Dioxane							
THF						0.04	V4/160
Esters							
Me acetate							
Et acetate							
i-Propyl acetate							
n-Butyl acetate							
Cellosolve acetate							
Miscellaneous							
DMF							
DMAc							
DMSO							
Sulfolane							
CS_2							
Acetic acid							
Aniline							
Nitrobenzene							
Morpholine							
Pyridine							
2-Nitropropane							
Acetonitrile							
Furfuraldehyde						0.04	V4/159
Phenol							
Water	96	65					

Trichloroethylene

Alternative names
1,2,2-trichloroethylene, trike, TCE, triclene, trilane, trichloroethene, **not** trichlorethane

Reference codes

CAS number	79 01 6	Hazchem code	2Z
UN number	1710	EPA code	U228

Physical properties

Molecular weight	131	Cubic expansion coeff (per °C × 10^3)	1.17
Empirical formula	$C_2H_1Cl_3$	Surface tension (@20°C dyn/cm)	29.5
Boiling point (°C)	87	Absolute viscosity (@25°C cP)	0.57
Freezing point (°C)	–86	Refractive index (25°C)	1.475
Specific gravity (20/4)	1.464		

Fire hazards

Flash point (closed cup °C)	32*	Lower explosive limit (ppm)	80000
Autoignition temperature (°C)	420	Upper explosive limit (ppm)	105000
Electrical conductivity	8E–12		

Health hazards

IDLH (ppm)	500	Vapour concentration @21°C ppm	80260
OES-TWA	100	Vapour density (relative to air)	4.55
OES-STEL	150	Vapour pressure @21°C mmHg	56.5
Odour threshold (ppm)	200	POCP	6.6

Aqueous effluent

Solubility in water (25°C %w/w)	0.11
Solubility of water in (25°C %w/w)	0.033
Log_{10} activated carbon partition	5.0
Log_{10} partition in octanol/water (w/w)	+2.29
Biological oxygen demand w/w (days)	
Theoretical oxygen demand w/w	0.61

Vapour pressure equation constants (Log_{10}, mmHg)

Antoine equation	A	6.51827
	B	1018.603
	C	192.731
Cox chart	A	
	B	

Solvent properties

Solubility parameter	8.0	Kauri butanol value	130
Dipole (D)	0.9	Evaporation time (ether = 1)	3.1
Dielectric constant (20°C)	3.42	Evaporation time (BuAc = 1)	4.9
Polarity (water 100)	16.0		

Thermal information

Latent heat (cal/mol)	7467
Nett heat of combustion (kcal/gmol)	206
Specific heat (cal/mol/°C)	30
Critical pressure (MN/m^2)	4.90
Critical temperature (K)	571
Latent heat of fusion (cal/mol)	
Van der Waals' volume	3.31
Van der Waals' surface area	2.86
Molar volume	90.01

*Very resistant to flashing.

Solute	Azeotrope X% w/w	°C	Reference	Solute γ^∞	Reference	Partition coefficient	Reference
Hydrocarbons							
n-Pentane	None		1482				
n-Hexane	None		2330	1.5	6a/463		
n-Heptane	None		2335				
n-Octane							
n-Nonane							
n-Decane							
2,2,4-TMP							
Cyclohexane	17	80	2328	1.3	6a/155		
Benzene	None		2326	1.0	7/114		
Toluene				0.8	7/370		
Ethylbenzene							
Xylenes							
C_9 Aromatics							
Tetralin							
Alcohols							
Methanol	62	59	1915	6.9	2a/40	8.9	V2/79
Ethanol	71	72	2286	7.3	2a/295	2.03	CEH
n-Propanol	83	82	2296	2.7	2a/518	0.24	V4/152
i-Propanol	70	75	2295	3.1	2d/43		
n-Butanol	97	87	2306	3.7	2f/121	0.08	V4/154
i-Butanol	91	85	2309				
s-Butanol	85	84	2307	3.8	2f/217		
n-Amyl alc.							
i-Amyl alc.	None		2321				
Cyclohexanol							
1-Octanol							
Ethanediol	None		2287				
DEG							
1,2-Propanediol							
Glycol ethers							
PGME							
EGME							
EEE							
EGBE							
Chlorinated							
MDC							
Chloroform				1.1	1x/3/943		
Carbon tet.	None		1093	1.0	1x/3/943		
1,2-EDC	33	82	2281	1.4	8/351		
1,1,1-TCA				1.0	1x/3/943		
TCE	–		–	–	–	–	–
Perk.				1.6	8/326		
MCB							

Chlorinated solvents

Solute	Azeotrope X% w/w	Azeotrope °C	Azeotrope Reference	Solute γ^∞	Reference	Partition coefficient	Reference
Ketones							
Acetone	None		2289	1.0	3b/51		
MEK	None		2299	1.3	3+4/264	0.05	V2/157
MIBK				1.1	3b/517		
Cyclohexanone							
NMP							
Acetophenone							
Ethers							
Diethyl ether							
DIPE	None		2331				
Dibutyl ether							
MTBE							
1,4-Dioxane	None		2301				
THF						0.05	V4/151
Esters							
Me acetate							
Et acetate	None		2302	1.0	5/454		
i-Propyl acetate	None		2317				
n-Butyl acetate				0.6	5/575		
Cellosolve acetate							
Miscellaneous							
DMF							
DMAc							
DMSO							
Sulfolane							
CS^2							
Acetic acid	96	86	2282	4.7	5/72	2.7	V2/151
Aniline							
Nitrobenzene							
Morpholine							
Pyridine							
2-Nitropropane							
Acetonitrile	71	75	2280	5.5	8/349		
Furfuraldehyde				3.1	3+4/37	0.03	V4/155
Phenol							
Water	94	73	218				

Perchloroethylene

Alternative names
Tetrachloroethylene, perk, tetrachloroethene

Reference codes

CAS number	127 18 4	Hazchem code	2Z
UN number	1897	EPA code	U210

Physical properties

Molecular weight	166	Cubic expansion coeff (per °C × 10^3)	1.02
Empirical formula	C_2Cl_4	Surface tension (@20°C dyn/cm)	32
Boiling point (°C)	122	Absolute viscosity (@25°C cP)	0.88
Freezing point (°C)	–36	Refractive index (25°C)	1.504
Specific gravity (20/4)	1.63		

Fire hazards

Flash point (closed cup °C)	None	Lower explosive limit (ppm)	None
Autoignition temperature (°C)	None	Upper explosive limit (ppm)	None
Electrical conductivity	5.5E–4		

Health hazards

IDLH (ppm)	400	Vapour concentration @21°C ppm	20600
OES-TWA	50	Vapour density (relative to air)	5.8
OES-STEL	150	Vapour pressure @21°C mmHg	15.4
Odour threshold (ppm)	300	POCP	0.5

Aqueous effluent

Solubility in water (25°C %w/w)	0.015
Solubility of water in (25°C %w/w)	0.0105
Log_{10} activated carbon partition	5.4
Log_{10} partition in octanol/water (w/w)	+2.60
Biological oxygen demand w/w (days)	0.06 (5)
Theoretical oxygen demand w/w	0.39

Vapour pressure equation constants (Log_{10}, mmHg)

Antoine equation	A	7.62930
	B	1803.96
	C	258.976
Cox chart	A	
	B	

Solvent properties

Solubility parameter	4.5	Kauri butanol value	90
Dipole (D)	0	Evaporation time (ether = 1)	6.0
Dielectric constant (20°C)	2.3	Evaporation time (BuAc = 1)	2.6
Polarity (water 100)			

Thermal information

Latent heat (cal/mol)	8316
Nett heat of combustion (kcal/gmol)	162
Specific heat (cal/mol/°C)	35
Critical pressure (MN/m^2)	4.48
Critical temperature (K)	613
Latent heat of fusion (cal/mol)	
Van der Waals' volume	3.89
Van der Waals' surface area	3.40
Molar volume	101.84

Solute	Azeotrope X% w/w	°C	Reference	Solute γ^{∞}	Reference	Partition coefficient	Reference
Hydrocarbons							
n-Pentane							
n-Hexane	None		2217a	1.5	6a/453		
n-Heptane							
n-Octane	92	120	2227				
n-Nonane							
n-Decane							
2,2,4-TMP							
Cyclohexane							
Benzene				1.3	7/112		
Toluene	None		2220				
Ethylbenzene	None		2225				
Xylenes	None		1167c				
C_9 Aromatics							
Tetralin							
Alcohols							
Methanol	36	64	1914	16.4	2a/37		
Ethanol	37	77	2162	5.4	2c/285		
n-Propanol	30	82	2176				
i-Propanol	52	94	2177	5.9	2d/42	2.6	CEH
n-Butanol	71	109	2186	3.8	2d/155		
i-Butanol	60	103	2189			21.7	CEH
s-Butanol	43	97	2187	3.3	2d/240		
n-Amyl alc.	85	117	2201				
i-Amyl alc.	81	116	2203				
Cyclohexanol							
1-Octanol							
Ethanediol	94	119	2163				
DEG							
1,2-Propanediol							
Glycol ethers							
PGME	95	121	2205				
EGME	76	110	2178				
EEE	84	116	2190				
EGBE	None		2218				
Chlorinated							
MDC				1.1	8/256		
Chloroform	None		1431	1.2	8/215		
Carbon tet.	None		1091	1.0	8/78		
1,2-EDC				2.0	8/340		
1,1,1-TCA							
TCE				1.1	8/327		
Perk.	–		–	–	–	–	–
MCB							

Chlorinated solvents

Solute	Azeotrope X% w/w	°C	Reference	Solute γ^∞	Reference	Part coefficient	Reference
Ketones							
Acetone	None		2168	2.6	3b/49	4.2	CEH
MEK							
MIBK	48	114	2209				
Cyclohexanone							
NMP							
Acetophenone							
Ethers							
Diethyl ether							
DIPE							
Dibutyl ether							
MTBE							
1,4-Dioxane	None		2181				
THF					0.08	V4/149	
Esters							
Me acetate							
Et acetate							
i-Propyl acetate							
n-Butyl acetate	79	120	2210				
Cellosolve acetate							
Miscellaneous							
DMF							
DMAc							
DMSO							
Sulfolane							
CS_2				1.4	8.316		
Acetic acid	61	107	2158			7.25	V2/147
Aniline							
Nitrobenzene							
Morpholine							
Pyridine	52	113	2192	2.0	8/346		
2-Nitropropane							
Acetonitrile							
Furfuraldehyde	None		2191	4.3	3a/117	0.05	V4/150
Phenol							
Water	84	88	217				

Monochlorobenzene

Alternative names
Chlorobenzene, MCB, phenyl chloride

Reference codes

CAS number	108 90 7	Hazchem code	2Y
UN number	1134	EPA code	U037

Physical properties

Molecular weight	113	Cubic expansion coeff (per °C × 10^3)	0.98
Empirical formula	$C_6H_5Cl_1$	Surface tension (@20°C dyn/cm)	33
Boiling point (°C)	132	Absolute viscosity (@25°C cP)	0.8
Freezing point (°C)	–46	Refractive index (25°C)	1.523
Specific gravity (20/4)	1.106		

Fire hazards

Flash point (closed cup °C)	29	Lower explosive limit (ppm)	13000
Autoignition temperature (°C)	640	Upper explosive limit (ppm)	71000
Electrical conductivity	7E–11		

Health hazards

IDLH (ppm)	2400	Vapour concentration @21°C ppm	12650
OES-TWA	50	Vapour density (relative to air)	3.9
OES-STEL		Vapour Pressure @21°C mmHg	9.5
Odour threshold (ppm)	1	POCP	

Aqueous effluent

Solubility in water (25°C %w/w)	0.049
Solubility of water in (25°C %w/w)	0.033
Log_{10} activated carbon partition	4.9
Log_{10} partition in octanol/water (w/w)	+2.84
Biological oxygen demand w/w (days)	0.03 (5)
Theoretical oxygen demand w/w	2.05

Vapour pressure equation constants (Log_{10}, mmHg)

Antoine equation	A	7.17294
	B	1549.200
	C	229.260
Cox chart	A	7.18576
	B	1558.4

Solvent properties

Solubility parameter	9.5	Kauri butanol value	90
Dipole (D)	1.3	Evaporation time (ether = 1)	10.0
Dielectric constant (20°C)	5.62	Evaporation time (BuAc = 1)	
Polarity (water 100)	18.8		

Thermal information

Latent heat (cal/mol)	8814
Nett heat of combustion (kcal/gmol)	754
Specific heat (cal/mol/°C)	35
Critical pressure (MN/m^2)	4.52
Critical temperature (K)	632
Latent heat of fusion (cal/mol)	2305
Van der Waals' volume	3.81
Van der Waals' surface area	2.84
Molar volume	102.24

Solute	Azeotrope X% w/w	Azeotrope °C	Reference	Solute γ^∞	Reference	Partition coefficient	Reference
Hydrocarbons							
n-Pentane							
n-Hexane	None		10519	1.6	6a/529		
n-Heptane	None		10531	1.8	6b/119		
n-Octane	None		10536	2.2	1x/3/1175		
n-Nonane							
n-Decane				1.5	6b/392		
2,2,4-TMP							
Cyclohexane	None		10515	1.3	6a/202		
Benzene	None		10509	1.0	7/243		
Toluene	None		10524	1.0	7/416		
Ethylbenzene	None		10532	1.0	7/469		
Xylenes	None		10534	0.9	7/508		
C_9 Aromatics							
Tetralin							
Alcohols							
Methanol	None		2063	4.9	2a/204		
Ethanol	None		4070	4.2	2a/397		
n-Propanol	20	97	6489	3.0	2a/552		
i-Propanol	None		6273	3.4	2d/64		
n-Butanol	44	115	8133	2.5	2b175		
i-Butanol	37	107	8331	2.8	2d/357		
s-Butanol	None		8230	2.8	2b/258		
n-Amyl alc.	75	126	9747				
i-Amyl alc.	66	124	9819				
Cyclohexanol	None			2.6	2b/395		
1-Octanol							
Ethanediol	94	130	4233				
DEG							
1,2-Propanediol							
Glycol ethers							
PGME	None		9960				
EGME	53	119	6566	3.4	2d/120		
EEE	68	127	8423				
EGBE	None		10521			0.16	P3988
Chlorinated							
MDC							
Chloroform	None		1484	1.2	8/244		
Carbon tet.	None		1152	1.2	8/166		
1,2-EDC							
1,1,1-TCA							
TCE							
Perk.							
MCB	–		–	–	–	–	–

Chlorinated solvents

Solute	Azeotrope X% w/w	°C	Reference	Solute γ^∞	Reference	Partition coefficient	Reference
Ketones							
Acetone	None		5372	1.3	3+4/192	0.17	V2/477
MEK	None			1.3	3+4/283	0.08	V3/22
MIBK	None			1.1	3b/543		
Cyclohexanone	None		10512				
NMP							
Acetophenone							
Ethers							
Diethyl ether							
DIPE							
Dibutyl ether	None		10537				
MTBE							
1,4-Dioxane				0.9	1x/3/1175		
THF							
Esters							
Me acetate				1.2	5/374		
Et acetate	None		7578	1.4	5/492		
i-Propyl acetate							
n-Butyl acetate	None		10516				
Cellosolve acetate							
Miscellaneous							
DMF							
DMAc							
DMSO							
Sulfolane							
CS_2							
Acetic acid	41	115	3160			3.7	CEH
Aniline	None		10511	1.8	8/527		
Nitrobenzene	None		10508	1.1	1x/3/1175		
Morpholine							
Pyridine						0.08	V3/226
2-Nitropropane							
Acetonitrile	None		2794a	3.3	8/381	0.32	V2/176
Furfuraldehyde	None		8758			0.01	V4/255
Phenol	None		10510				
Water	72	90	484				

Section 5

Ketones

Acetone

Alternative names
Propan-2-one, dimethyl ketone

Reference codes

CAS number	67 64 1	Hazchem code	2YE
UN number	1090	EPA code	U002

Physical properties

Molecular weight	58	Cubic expansion coeff (per °C × 10^3)	1.4
Empirical formula	$C_3H_6O_1$	Surface tension (@20°C dyn/cm)	23.3
Boiling point (°C)	56	Absolute viscosity (@25°C cP)	0.33
Freezing point (°C)	–95	Refractive index (25°C)	1.357
Specific gravity (20/4)	0.790		

Fire hazards

Flash point (closed cup °C)	–18	Lower explosive limit (ppm)	26000
Autoignition temperature (°C)	465	Upper explosive limit (ppm)	128000
Electrical conductivity	5E–9		

Health hazards

IDLH (ppm)	20000	Vapour concentration @21°C ppm	342800
OES-TWA	750	Vapour density (relative to air)	2.0
OES-STEL	1500	Vapour pressure @21°C mmHg	194
Odour threshold (ppm)	300	POCP	17.8

Aqueous effluent

Solubility in water (25°C %w/w)	Total	
Solubility of water in (25°C %w/w)	Total	
Log_{10} activated carbon partition	1.74	
Log_{10} partition in octanol/water (w/w)	–0.24	
Biological oxygen demand w/w (days)		1.22 (5)
Theoretical oxygen demand w/w	2.21	

Vapour pressure equation constants (Log_{10}, mmHg)

Antoine equation	A	7.11714
	B	1210.596
	C	229.664
Cox chart	A	7.18990
	B	1232.4

Solvent properties

Solubility parameter	10.0	Kauri butanol value	
Dipole (D)	2.9	Evaporation time (ether = 1)	1.8
Dielectric constant (20°C)	20.6	Evaporation time (BuAc = 1)	5.6
Polarity (water 100)	35.5		

Thermal information

Latent heat (cal/mol)	7076
Nett heat of combustion (kcal/gmol)	395
Specific heat (cal/mol/°C)	30
Critical pressure (MN/m^2)	4.8
Critical temperature (K)	508
Latent heat of fusion (cal/mol)	1358
Van der Waals' volume	2.57
Van der Waals' surface area	2.34
Molar volume	73.4

Solute	Azeotrope X% w/w	Azeotrope °C	Reference	Solute γ^∞	Reference	Partition coefficient	Reference
Hydrocarbons							
n-Pentane	21	32	5368	5.8	3+4/190		
n-Hexane	59	50	5385	5.1	3+4/225		
n-Heptane	90	56	5393	5.5	3+4/242		
n-Octane	None		5395	8.4	3b/224		
n-Nonane	None			6.2	3b/236		
n-Decane	None		5396	6.7	3+4/247		
2,2,4-TMP	None			8.2	3b/225		
Cyclohexane	67	53	5378	4.3	3+4/213		
Benzene	None		5374	1.4	3+4/195		
Toluene	None		5391	1.6	3+4/236		
Ethylbenzene	None			2.1	3b/217		
Xylenes	None			2.6	3b/222		
C_9 Aromatics	None			1.6	3b/233		
Tetralin							
Alcohols							
Methanol	88	55	1963	1.8	2a/68		
Ethanol	None		3965	1.7	2a/321		
n-Propanol	None		5320				
i-Propanol	None		5319	2.4	2b/43		
n-Butanol	None		5344	1.6	2b/140		
i-Butanol	None		5347	1.2	2f/304		
s-Butanol							
n-Amyl alc.							
i-Amyl alc.							
Cyclohexanol	None			4.6	2d/510		
1-Octanol							
Ethanediol							
DEG							
1,2-Propanediol							
Glycol ethers							
PGME							
EGME				2.9	2d/113		
EEE	None			1.0	2f/332		
EGBE							
Chlorinated							
MDC	None		1553	0.7	3b/27		
Chloroform	22	64	1443	0.6	3+4/90		
Carbon tet.	89	56	1108	2.3	3+4/80		
1,2-EDC	None		2966	1.0	3+4/144		
1,1,1-TCA							
TCE	None		2289	2.7	3b/51		
Perk.	None		2168	3.7	3b/49		
MCB	None		5372	1.6	3+4/192		

Ketones

Solute	Azeotrope X% w/w	°C	Reference	Solute γ^∞	Reference	Partition coefficient	Reference
Ketones							
Acetone	–		–	–	–	–	–
MEK	None		5330	0.9	3+4/173		
MIBK	None		5382	1.1	3b/196		
Cyclohexanone							
NMP							
Acetophenone							
Ethers							
Diethyl ether	None		5346	1.4	3+4/177		
DIPE	61	54	5386				
Dibutyl ether							
MTBE							
1,4-Dioxane	None		5333	1.4	1x/3/991		
THF							
Esters							
Me acetate	50	55	5310	1.3	3+4/159		
Et acetate	None		5334	1.2	3+4/176		
i-Propyl acetate	None		5365				
n-Butyl acetate	None		5383	1.1	3b/197		
Cellosolve acetate							
Miscellaneous							
DMF	None			3.1	3+4/164		
DMAc							
DMSO	None			1.2	3b/80		
Sulfolane							
CS_2	33	39	1194	3.6	3+4/132		
Acetic acid	None		3101	1.4	3+4/148		
Aniline	None			0.9	3b/183		
Nitrobenzene							
Morpholine							
Pyridine	None		5353	2.1	3+4/181		
2-Nitropropane							
Acetonitrile	None		2762	1.0	3+4/143		
Furfuraldehyde							
Phenol	None		5375				
Water	None		269	5.3	1x/3/993		

Methyl ethyl ketone

Alternative names
MEK, butan-2-one

Reference codes

CAS number	78 93 3	Hazchem code	2YE
UN number	1193	EPA code	U159

Physical properties

Molecular weight	72	Cubic expansion coeff (per °C × 10^3)	1.3
Empirical formula	$C_4H_8O_1$	Surface tension (@20°C dyn/cm)	24.6
Boiling point (°C)	80	Absolute viscosity (@25°C cP)	0.41
Freezing point (°C)	–87	Refractive index (25°C)	1.377
Specific gravity (20/4)	0.805		

Fire hazards

Flash point (closed cup °C)	–6	Lower explosive limit (ppm)	18000
Autoignition temperature (°C)	485	Upper explosive limit (ppm)	100000
Electrical conductivity	3.6E–9		

Health hazards

IDLH (ppm)	3000	Vapour concentration @21°C ppm	112000
OES-TWA	200	Vapour density (relative to air)	2.50
OES-STEL	300	Vapour pressure @21°C mmHg	75.3
Odour threshold (ppm)	30	POCP	42.3

Aqueous effluent

Solubility in water (25°C %w/w)	26
Solubility of water in (25°C %w/w)	12.0
Log_{10} activated carbon partition	2.25
Log_{10} partition in octanol/water (w/w)	+0.29
Biological oxygen demand w/w (days)	2.14 (5)
Theoretical oxygen demand w/w	2.44

Vapour pressure equation constants (Log_{10}, mmHg)

Antoine equation	A	7.06356
	B	1261.340
	C	221.969
Cox chart	A	7.22242
	B	1345.9

Solvent properties

Solubility parameter	9.3	Kauri butanol value	
Dipole (D)	2.8	Evaporation time (ether = 1)	2.5
Dielectric constant (20°C)	18.5	Evaporation time (BuAc = 1)	4.6
Polarity (water 100)	32.7		

Thermal information

Latent heat (cal/mol)	7848
Nett heat of combustion (kcal/gmol)	540
Specific heat (cal/mol/°C)	38
Critical pressure (MN/m^2)	4.16
Critical temperature (K)	535
Latent heat of fusion (cal/mol)	1790
Van der Waals' volume	3.25
Van der Waals' surface area	2.88
Molar volume	89.44

Solute	Azeotrope X% w/w	°C	Reference	Solute γ^∞	Reference	Partition coefficient	Reference
Hydrocarbons							
n-Pentane				3.4	1x/3/1044		
n-Hexane	29	64	7376	2.9	3+4/302		
n-Heptane	70	77	7384	3.2	3+4/311	0.12	CEH
n-Octane				6.0	3+4/317		
n-Nonane							
n-Decane	None			5.0	3b/396		
2,2,4-TMP	None			4.3	3b/395		
Cyclohexane	40	72	7374	3.3	3+4/297		
Benzene	45	78	7369	1.2	3+4/284		
Toluene	None		7382	1.6	3+4/308		
Ethylbenzene	None			1.9	3+4/316		
Xylenes				1.3	3b/382		
C_9 Aromatics							
Tetralin							
Alcohols							
Methanol	30	64	1993	2.0	2a/133		
Ethanol	61	74	4005	2.3	2a/343		
n-Propanol	None		6445	1.6	2c/496	0.25	V3/225
i-Propanol	68	78	6335	1.6	2b/54		
n-Butanol	None		7357	0.9	2f/144	0.22	V3/17
i-Butanol	None		7360				
s-Butanol	None		7358	1.3	2b/239		
n-Amyl alc.							
i-Amyl alc.							
Cyclohexanol							
1-Octanol							
Ethanediol						8.16	V2/418
DEG							
1,2-Propanediol							
Glycol ethers							
PGME							
EGME				1.8	2b/122		
EEE				1.6	2f/334		
EGBE	None			2.4	2b/430	0.08	CEH
Chlorinated							
MDC	None		1564	0.6	3+4//261		
Chloroform	83	80	1460	0.5	3+4/260		
Carbon tet.	29	74	1121	1.5	3+4/259		
1,2-EDC	24		2977	0.7	3b/271		
1,1,1-TCA							
TCE	None		2299	1.1	3+4/264		
Perk.				2.5	3b/265		
MCB	None			1.5	3+4/283		

Ketones

	Azeotrope						
Solute	X% w/w	°C	Reference	Solute γ^∞	Reference	Partition coefficient	Reference
Ketones							
Acetone	None		5330	0.9	3+4/173	0.30	V4/215
MEK	–		–	–	–	–	–
MIBK				1.0	3+4/300		
Cyclohexanone							
NMP							
Acetophenone							
Ethers							
Diethyl ether							
DIPE				1.5	3b/357		
Dibutyl ether							
MTBE							
1,4-Dioxane	None		7344	1.2	1x/3/1044		
THF							
Esters							
Me acetate	None		5519	1.0	3+4/271		
Et acetate	12	77	7345	1.1	3+4/278		
i-Propyl acetate	None		7366				
n-Butyl acetate							
Cellosolve acetate							
Miscellaneous							
DMF				1.4	3b/289		
DMAc							
DMSO							
Sulfolane							
CS_2	16	46	1216	3.0	1x/1/97		
Acetic acid	None		3117	1.6	3+4/269	0.21	P326
Aniline							
Nitrobenzene				1.5	3b/316		
Morpholine							
Pyridine				1.1	1x/1/98		
2-Nitropropane	None		6272				
Acetonitrile	73			1.2	3b/268		
Furfuraldehyde							
Phenol	None		7370	0.1	2b/358		
Water	89	73	338	6.9	1x/1/99		

Methyl isobutyl ketone

Alternative names
4-Methyl-2-pentanone, MIBK

Reference codes

CAS number	108 10 1	Hazchem code	
UN number	1245	EPA code	U161

Physical properties

Molecular weight	100	Cubic expansion coeff (per °C × 10^3)	0.94
Empirical formula	$C_6H_{12}O_1$	Surface tension (@20°C dyn/cm)	23.6
Boiling point (°C)	116	Absolute viscosity (@25°C cP)	0.61
Freezing point (°C)	–84	Refractive index (25°C)	1.394
Specific gravity (20/4)	0.801		

Fire hazards

Flash point (closed cup °C)	13	Lower explosive limit (ppm)	14000
Autoignition temperature (°C)	459	Upper explosive limit (ppm)	75000
Electrical conductivity	5E–8		

Health hazards

IDLH (ppm)		Vapour concentration @21°C ppm	21700
OES-TWA	50	Vapour density (relative to air)	3.47
OES-STEL		Vapour pressure @21°C mmHg	16.5
Odour threshold (ppm)	8	POCP	63.3

Aqueous effluent

Solubility in water (25°C %w/w)	1.7
Solubility of water in (25°C %w/w)	1.9
Log_{10} activated carbon partition	3.05
Log_{10} partition in octanol/water (w/w)	
Biological oxygen demand w/w (days)	2.06
Theoretical oxygen demand w/w	2.2

Vapour pressure equation constants (Log_{10}, mmHg)

Antoine equation	A	6.67272
	B	1168.408
	C	191.944
Cox chart	A	7.27155
	B	1519.2

Solvent properties

Solubility parameter	8.4	Kauri butanol value	
Dipole (D)	2.81	Evaporation time (ether = 1)	5.6
Dielectric constant (20°C)	13.1	Evaporation time (BuAc = 1)	1.4
Polarity (water 100)	27		

Thermal information

Latent heat (cal/mol)	8500
Nett heat of combustion (kcal/gmol)	672
Specific heat (cal/mol/°C)	46
Critical pressure (MN/m^2)	3.27
Critical temperature (K)	571.5
Latent heat of fusion (cal/mol)	
Van der Waals' volume	4.60
Van der Waals' surface area	3.95
Molar volume	125.8

Solute	Azeotrope X% w/w	°C	Reference	Solute γ^∞	Reference	Partition coefficient	Reference
Hydrocarbons							
n-Pentane				2.3	1x/3/1233		
n-Hexane				2.5	1x/1/273		
n-Heptane	13	98	11801	2.1	3b/550		
n-Octane	65	113	11805	2.0	1x/1/273		
n-Nonane							
n-Decane							
2,2,4-TMP				1.9	1x/1/373		
Cyclohexane	None		11685	1.2	3+4/354		
Benzene	None		10857	1.0	3+4/351		
Toluene	3	111	11799	1.1	3+4/356		
Ethylbenzene	None		11802				
Xylenes	None			1.6	3b/553		
C_9 Aromatics							
Tetralin							
Alcohols							
Methanol	None		2084	2.1	2a/248		
Ethanol	None		4101	2.5	2c/423		
n-Propanol							
i-Propanol	None		6386	1.5	2b/96		
n-Butanol	70	114	8152	2.5	2b/193		
i-Butanol	9	108	8343				
s-Butanol							
n-Amyl alc.				30.2	2f/380		
i-Amyl alc.	None		9836				
Cyclohexanol							
1-Octanol							
Ethanediol						8.65	V2/430
DEG							
1,2-Propanediol							
Glycol ethers							
PGME							
EGME	75	114	6575				
EEE	None		8433				
EGBE							
Chlorinated							
MDC							
Chloroform	None		1492	0.5	3+4/343		
Carbon tet.				1.1	1x/3/1233		
1,2-EDC				0.8	3b/519		
1,1,1-TCA				0.9	1x/3/1233		
TCE				1.1	3b/517		
Perk.	52	114	2209				
MCB	None			1.0	3b/543		

Ketones

Solute	Azeotrope X% w/w	°C	Reference	Solute γ^∞	Reference	Partition coefficient	Reference
Ketones							
Acetone	None		5382	1.2	1x/1/273	0.12	V2/485
MEK				1.1	3+4/300		
MIBK	–		–	–	–	–	–
Cyclohexanone							
NMP							
Acetophenone							
Ethers							
Diethyl ether							
DIPE							
Dibutyl ether							
MTBE							
1,4-Dioxane				1.2	3b/523		
THF							
Esters							
Me acetate							
Et acetate				1.9	3b/527		
i-Propyl acetate							
n-Butyl acetate							
Cellosolve acetate							
Miscellaneous							
DMF							
DMAc							
DMSO							
Sulfolane							
CS_2	None		1272				
Acetic acid	None		3175	1.8	3+4/345	0.32	V2/285
Aniline							
Nitrobenzene							
Morpholine							
Pyridine	40	115	8849	1.0	3b/531		
2-Nitropropane							
Acetonitrile						0.08	V2/183
Furfuraldehyde				1.4	3a/126	0.03	V3/193
Phenol						<0.01	V3/295
Water	76	88	537	10.6	1b/337		

Cyclohexanone

Alternative names
Cyclohexyl ketone, sextone

Reference codes

CAS number	108 94 1	Hazchem code	3Y
UN number	1915	EPA code	U057

Physical properties

Molecular weight	98	Cubic expansion coeff (per °C × 10^3)	0.94
Empirical formula	$C_6H_{10}O_1$	Surface tension (@20°C dyn/cm)	34.5
Boiling point (°C)	156	Absolute viscosity (@25°C cP)	2.2
Freezing point (°C)	–32	Refractive index (25°C)	1.448
Specific gravity (20/4)	0.948		

Fire hazards

Flash point (closed cup °C)	43	Lower explosive limit (ppm)	11000
Autoignition temperature (°C)	420	Upper explosive limit (ppm)	94000
Electrical conductivity	5E–18		

Health hazards

IDLH (ppm)	5000	Vapour concentration @21°C ppm	3963
OES-TWA	25	Vapour density (relative to air)	3.40
OES-STEL	100	Vapour pressure @21°C mmHg	3.1
Odour threshold (ppm)	1	POCP	

Aqueous effluent

Solubility in water (25°C %w/w)	2.3
Solubility of water in (25°C %w/w)	8.0
Log_{10} activated carbon partition	3.0
Log_{10} partition in octanol/water (w/w)	+0.81
Biological oxygen demand w/w (days)	1.23
Theoretical oxygen demand w/w	2.61

Vapour pressure equation constants (Log_{10}, mmHg)

Antoine equation	A	7.47050
	B	1832.200
	C	244.200
Cox chart	A	7.32768
	B	1716.5

Solvent properties

Solubility parameter	9.9	Kauri butanol value	
Dipole (D)	3.1	Evaporation time (ether = 1)	41
Dielectric constant (20°C)	18.2	Evaporation time (BuAc = 1)	0.25
Polarity (water 100)	28		

Thermal information

Latent heat (cal/mol)	9016
Nett heat of combustion (kcal/gmol)	788
Specific heat (cal/mol/°C)	48
Critical pressure (MN/m^2)	3.8
Critical temperature (K)	629
Latent heat of fusion (cal/mol)	
Van der Waals' volume	4.14
Van der Waals' surface area	3.34
Molar volume	104.2

Solute	Azeotrope X% w/w	°C	Reference	Solute γ^{∞}	Reference	Partition coefficient	Reference
Hydrocarbons							
n-Pentane				3.7	1x/1/257		
n-Hexane				4.1	1x/1/258		
n-Heptane				4.4	3b/509		
n-Octane				4.9	1x/3/1210		
n-Nonane							
n-Decane							
2,2,4-TMP							
Cyclohexane				1.6	3b/505		
Benzene				0.9	3b/503		
Toluene	None			1.0	3+4/339		
Ethylbenzene							
Xylenes	None		11367	1.5	3b/511		
C_9 Aromatics							
Tetralin							
Alcohols							
Methanol							
Ethanol				2.1	1x/3/1210		
n-Propanol							
i-Propanol							
n-Butanol							
i-Butanol							
s-Butanol							
n-Amyl alc.							
i-Amyl alc.							
Cyclohexanol	None		11357				
1-Octanol							
Ethanediol						3.4	V2/428
DEG							
1,2-Propanediol							
Glycol ethers							
PGME							
EGME							
EEE							
EGBE						0.03	P3981
Chlorinated							
MDC				0.4	1x/1/256		
Chloroform				0.3	1x/1/256		
Carbon tet.				1.2	3b/495		
1,2-EDC							
1,1,1-TCA							
TCE							
Perk.							
MCB	None		10512				

Ketones

Solute	Azeotrope X% w/w	°C	Reference	Solute γ^{∞}	Reference	Partition coefficient	Reference
Ketones							
Acetone				1.3	1x/1/256		
MEK				1.0	1x/3/1210		
MIBK							
Cyclohexanone	–		–	–	–	–	–
NMP							
Acetophenone							
Ethers							
Diethyl ether							
DIPE				2.5	3b/506		
Dibutyl ether							
MTBE							
1,4-Dioxane				1.0	1x/3/1210		
THF							
Esters							
Me acetate							
Et acetate				1.3	1x/1/256		
i-Propyl acetate							
n-Butyl acetate							
Cellosolve acetate							
Miscellaneous							
DMF				1.8	3b/500		
DMAc							
DMSO							
Sulfolane							
CS_2				1.9	1x/1/356		
Acetic acid							
Aniline	None		11144				
Nitrobenzene							
Morpholine							
Pyridine							
2-Nitropropane							
Acetonitrile				1.4	1x/1/256		
Furfuraldehyde	None		8762				
Phenol	28	185	10889	0.1	2b/368		
Water	43	96	506	5.93	1/511		

n-Methyl-2-pyrrolidone

Alternative names
M-pyrol, NMP, 1-methyl pyrrolidone

Reference codes

CAS number	872 50 4	Hazchem code	
UN number		EPA code	

Physical properties

Molecular weight	99	Cubic expansion coeff (per °C × 10^3)	0.9
Empirical formula	$C_5H_9N_1O_1$	Surface tension (@20°C dyn/cm)	40.7
Boiling point (°C)	202	Absolute viscosity (@25°C cP)	1.8
Freezing point (°C)	–24	Refractive index (25°C)	1.468
Specific gravity (20/4)	1.03		

Fire hazards

Flash point (closed cup °C)	95	Lower explosive limit (ppm)	21800
Autoignition temperature (°C)	287	Upper explosive limit (ppm)	122400
Electrical conductivity	2E–8		

Health hazards

IDLH (ppm)		Vapour concentration @21°C ppm	395
OES-TWA	100	Vapour density (relative to air)	3.44
OES-STEL		Vapour pressure @21°C mmHg	0.3
Odour threshold (ppm)		POCP	

Aqueous effluent

Solubility in water (25°C %w/w)	Total
Solubility of water in (25°C %w/w)	Total
Log_{10} activated carbon partition	
Log_{10} partition in octanol/water (w/w)	
Biological oxygen demand w/w (days)	1.1 (5)
Theoretical oxygen demand w/w	2.18

Vapour pressure equation constants (Log_{10}, mmHg)

Antoine equation	A	8.27890
	B	2570.30
	C	273.150
Cox chart	A	
	B	

Solvent properties

Solubility parameter	11.0	Kauri butanol value	
Dipole (D)	4.1	Evaporation time (ether = 1)	
Dielectric constant (20°C)	32.2	Evaporation time (BuAc = 1)	0.04
Polarity (water 100)	36		

Thermal information

Latent heat (cal/mol)	12600
Nett heat of combustion (kcal/gmol)	667
Specific heat (cal/mol/°C)	40
Critical pressure (MN/m^2)	
Critical temperature (K)	722
Latent heat of fusion (cal/mol)	6207
Van der Waals' volume	3.98
Van der Waals' surface area	3.20
Molar volume	96.1

Solute	Azeotrope X% w/w	Azeotrope °C	Azeotrope Reference	Solute γ^∞	Reference	Partition coefficient	Reference
Hydrocarbons							
n-Pentane				10.2	1x/1/171		
n-Hexane				11.3	1x/1/174		
n-Heptane				10.5	1x/1/176		
n-Octane				12.1	1x/1/176		
n-Nonane				13.9	1x/1/177		
n-Decane				15.9	1x/1/177		
2,2,4-TMP				14.4	1x/1/177		
Cyclohexane				5.9	1x/1/173		
Benzene				1.2	1x/1/172		
Toluene				1.5	1x/1/175		
Ethylbenzene				1.8	1x/1/176		
Xylenes				1.8	1x/1/1125		
C_9 Aromatics				2.0	3b/463		
Tetralin							
Alcohols							
Methanol				0.5	1x/1/169		
Ethanol				0.6	1x/1/169		
n-Propanol				0.7	1x/3/1118		
i-Propanol				0.75	1x/1/169		
n-Butanol				0.7	1x/3/1119		
i-Butanol				0.7	1x/3/1119		
s-Butanol							
n-Amyl alc.							
i-Amyl alc.							
Cyclohexanol				0.6	2f/411		
1-Octanol							
Ethanediol				0.05	2f/17		
DEG							
1,2-Propanediol							
Glycol ethers							
PGME							
EGME							
EEE							
EGBE							
Chlorinated							
MDC				0.4	1x/3/1117		
Chloroform				0.4	1x/3/1117		
Carbon tet.				2.3	1x/3/1117		
1,2-EDC							
1,1,1-TCA							
TCE							
Perk.							
MCB							

Ketones

Solute	Azeotrope X% w/w °C	Reference	Solute γ^{∞}	Reference	Partition coefficient	Reference
Ketones						
Acetone			1.3	1x/1/169		
MEK			1.4	1x/1/170		
MIBK						
Cyclohexanone						
NMP	–	–	–	–	–	–
Acetophenone						
Ethers						
Diethyl ether						
DIPE						
Dibutyl ether						
MTBE						
1,4-Dioxane			1.2	1x/3/1119		
THF						
Esters						
Me acetate			1.6	1x/1/169		
Et acetate			1.9	1x/3/1119		
i-Propyl acetate						
n-Butyl acetate						
Cellosolve acetate						
Miscellaneous						
DMF						
DMAc						
DMSO						
Sulfolane						
CS_2						
Acetic acid						
Aniline						
Nitrobenzene						
Morpholine						
Pyridine						
2-Nitropropane						
Acetonitrile						
Furfuraldehyde						
Phenol						
Water	None	416a	1.15	1a/379		

Acetophenone

Alternative names
Acetylbenzene, methyl phenyl ketone

Reference codes

CAS number	98 86 2	Hazchem code	
UN number		EPA code	

Physical properties

Molecular weight	120	Cubic expansion coeff (per °C × 10^3)	0.84
Empirical formula	$C_8H_8O_1$	Surface tension (@30°C dyn/cm)	12
Boiling point (°C)	202	Absolute viscosity (@25°C cP)	1.74
Freezing point (°C)	+19.6	Refractive index (25°C)	1.532
Specific gravity (20/4)	1.024		

Fire hazards

Flash point (closed cup °C)	82	Lower explosive limit (ppm)	
Autoignition temperature (°C)	570	Upper explosive limit (ppm)	
Electrical conductivity	3E–9		

Health hazards

IDLH (ppm)	1.0	Vapour concentration @21°C ppm	461
OES-TWA		Vapour density (relative to air)	4.17
OES-STEL		Vapour pressure @21°C mmHg	0.35
Odour threshold (ppm)	10	POCP	

Aqueous effluent

Solubility in water (25°C %w/w)	0.55
Solubility of water in (25°C %w/w)	1.70
Log_{10} activated carbon partition	3.84
Log_{10} partition in octanol/water (w/w)	
Biological oxygen demand w/w (days)	
Theoretical oxygen demand w/w	2.53

Vapour pressure equation constants (Log_{10}, mmHg)

Antoine equation	A	7.2273
	B	1774.6
	C	206.3
Cox chart	A	
	B	

Solvent properties

Solubility parameter		Kauri butanol value	
Dipole (D)	2.9	Evaporation time (ether = 1)	
Dielectric constant (20°C)	17.4	Evaporation time (BuAc = 1)	0.03
Polarity (water 100)	30.6		

Thermal information

Latent heat (cal/mol)	10032
Nett heat of combustion (kcal/gmol)	949
Specific heat (cal/mol/°C)	54
Critical pressure (MN/m^2)	3.8
Critical temperature (K)	428
Latent heat of fusion (cal/mol)	
Van der Waals' volume	4.69
Van der Waals' surface area	3.61
Molar volume	117.4

Solute	Azeotrope X% w/w	°C	Reference	Solute γ^{∞}	Reference	Partition coefficient	Reference
Hydrocarbons							
n-Pentane				5.4	1x/1/363		
n-Hexane				6.4	1x/1/363		
n-Heptane				6.0	1x/1/363		
n-Octane				7.6	1x/3/1345		
n-Nonane				7.3	1x/1/364		
n-Decane							
2,2,4-TMP							
Cyclohexane				4.7	1x/1/363		
Benzene				1.6	1x/1/363		
Toluene				1.3	1x/3/1345		
Ethylbenzene				1.7	1x/1/363		
Xylenes				1.8	1x/1/364		
C_9 Aromatics							
Tetralin							
Alcohols							
Methanol				3.5	1x/1/362		
Ethanol				3.3	1x/3/1345		
n-Propanol							
i-Propanol							
n-Butanol							
i-Butanol							
s-Butanol	None			1.1	2b/251		
n-Amyl alc.							
i-Amyl alc.							
Cyclohexanol							
1-Octanol	13	195	13909				
Ethanediol	48	186	4316				
DEG	None		8538				
1,2-Propanediol		184	6664				
Glycol ethers							
PGME							
EGME							
EEE							
EGBE							
Chlorinated							
MDC				0.6	1x/1/362		
Chloroform				0.6	1x/1/362		
Carbon tet.				1.7	1x/1/362		
1,2-EDC							
1,1,1-TCA							
TCE							
Perk.							
MCB							

Ketones

Solute	Azeotrope X% w/w	°C)	Reference	Solute γ^∞	Reference	Partition coefficient	Reference
Ketones							
Acetone							
MEK				1.00	1x/3/1345		
MIBK							
Cyclohexanone							
NMP							
Acetophenone	–		–	–	–	–	–
Ethers							
Diethyl ether							
DIPE							
Dibutyl ether							
MTBE							
1,4-Dioxane				0.9	1x/3/1345		
THF							
Esters							
Me acetate							
Et acetate							
i-Propyl acetate							
n-Butyl acetate							
Cellosolve acetate							
Miscellaneous							
DMF							
DMAc							
DMSO							
Sulfolane							
CS_2				2.1	1x/1/362		
Acetic acid							
Aniline							
Nitrobenzene	None		10733				
Morpholine							
Pyridine							
2-Nitropropane							
Acetonitrile				1.7	1x/1/362		
Furfuraldehyde						0.02	V3/205
Phenol				0.5	2f/402		
Water	18	98	671	6.44	1a/460		

Section 6

Ethers

Diethyl ether

Alternative names
Ethyl ether, ethoxy ethane, ether, ethyl oxide, sulfuric ether, **not** petroleum ether

Reference codes

CAS number	60 29 7	Hazchem code	3YE
UN number	1155	EPA code	U117

Physical properties

Molecular weight	74	Cubic expansion coeff (per °C × 10^3)	1.6
Empirical formula	$C_4H_{10}O_1$	Surface tension (@20°C dyn/cm)	17
Boiling point (°C)	34.5	Absolute viscosity (@25°C cP)	0.24
Freezing point (°C)	–116	Refractive index (25°C)	1.352
Specific gravity (20/4)	0.715		

Fire hazards

Flash point (closed cup °C)	–45	Lower explosive limit (ppm)	18500
Autoignition temperature (°C)	160	Upper explosive limit (ppm)	360000
Electrical conductivity	3E–16		

Health hazards

IDLH (ppm)	19000	Vapour concentration @21°C ppm	610000
OES-TWA	400	Vapour density (relative to air)	2.57
OES-STEL	500	Vapour pressure @21°C mmHg	462
Odour threshold (ppm)	1	POCP	60

Aqueous effluent

Solubility in water (25°C %w/w)	6.9
Solubility of water in (25°C %w/w)	1.3
Log_{10} activated carbon partition	
Log_{10} partition in octanol/water (w/w)	+0.77
Biological oxygen demand w/w (days)	0.03 (5)
Theoretical oxygen demand w/w	2.59

Vapour pressure equation constants (Log_{10}, mmHg)

Antoine equation	A	6.98472
	B	1090.64
	C	231.20
Cox chart	A	7.00353
	B	1088.4

Solvent properties

Solubility parameter	7.4	Kauri butanol value	
Dipole (D)	1.3	Evaporation time (ether = 1)	1.0
Dielectric constant (20°C)	4.3	Evaporation time (BuAc = 1)	28.0
Polarity (water 100)	11.7		

Thermal information

Latent heat (cal/mol)	6216
Nett heat of combustion (kcal/gmol)	598
Specific heat (cal/mol/°C)	40
Critical pressure (MN/m^2)	3.61
Critical temperature (K)	473
Latent heat of fusion (cal/mol)	1735
Van der Waals' volume	3.39
Van der Waals' surface area	3.02
Molar volume	103.5

Solute	Azeotrope X% w/w	Azeotrope °C	Reference	Solute γ^∞	Reference	Partition coefficient	Reference
Hydrocarbons							
n-Pentane	68	33	8296				
n-Hexane	None		8301				
n-Heptane							
n-Octane				1.5	1x/3/1077		
n-Nonane							
n-Decane							
2,2,4-TMP							
Cyclohexane							
Benzene	None		8293	0.9	3+4/516		
Toluene	None		8304	1.2	1x/3/1077		
Ethylbenzene							
Xylenes							
C_9 Aromatics							
Tetralin							
Alcohols							
Methanol	2	62		4.8	2a/170	2.9	P157
Ethanol	None		4029	3.8	2a/375	1.75	V2/341
n-Propanol	None		6464			0.18	P678
i-Propanol	None		6351			0.44	P649
n-Butanol	None		8104				
i-Butanol							
s-Butanol							
n-Amyl alc.							
i-Amyl alc.						1.3	P1274
Cyclohexanol							
1-Octanol							
Ethanediol							
DEG						55.7	P1004
1,2-Propanediol						13.4	P657
Glycol ethers							
PGME							
EGME						1.61	P655
EEE						1.22	P1001
EGBE							
Chlorinated							
MDC	30	41	1566	0.6	3+4/492		
Chloroform	None		1474	0.4	3+4/486		
Carbon tet.	None		1136				
1,2-EDC	None		2987				
1,1,1-TCA							
TCE							
Perk.							
MCB							

Ethers

Solute	Azeotrope X% w/w	°C	Reference	Solute γ^{∞}	Reference	Partition coefficient	Reference
Ketones							
Acetone	None		5346	2.2	3+4/177	0.28	V2/470
MEK				1.9	1x/3/1077		
MIBK							
Cyclohexanone							
NMP							
Acetophenone						<0.01	P2744
Ethers							
Diethyl ether	–		–	–	–	–	–
DIPE							
Dibutyl ether	None		8305				
MTBE							
1,4-Dioxane				2.0	1x/3/1077		
THF							
Esters							
Me acetate	None		5527			0.08	P516
Et acetate				1.1	3+4/513	0.03	P860
i-Propyl acetate							
n-Butyl acetate							
Cellosolve acetate							
Miscellaneous							
DMF						24.3	P594
DMAc							
DMSO							
Sulfolane							
CS_2	99	34	1235	1.6	3+4/495		
Acetic acid	None		3136	2.6	3+4/502	0.41	V2/228
Aniline						0.03	P1710
Nitrobenzene	None		8297				
Morpholine							
Pyridine						0.20	P1101
2-Nitropropane							
Acetonitrile	None		2778	2.5	3+4/499	0.40	P261
Furfuraldehyde							
Phenol						<0.01	P1617
Water	99	34	375	28.6	1a/257		

Diisopropyl ether

Alternative names
Isopropyl ether, DIPE

Reference codes			
CAS number	108 20 3	Hazchem code	3YE
UN number	1159	EPA code	

Physical properties			
Molecular weight	102	Cubic expansion coeff (per °C × 10^3)	1.4
Empirical formula	$C_6H_{14}O_1$	Surface tension (@20°C dyn/cm)	18
Boiling point (°C)	68	Absolute viscosity (@25°C cP)	0.33
Freezing point (°C)	–86	Refractive index (25°C)	1.367
Specific gravity (20/4)	0.724		

Fire hazards			
Flash point (closed cup °C)	–28	Lower explosive limit (ppm)	14000
Autoignition temperature (°C)	430	Upper explosive limit (ppm)	79000
Electrical conductivity			

Health hazards			
IDLH (ppm)	10000	Vapour concentration@ 21°C ppm	193000
OES-TWA	250	Vapour density (relative to air)	3.58
OES-STEL	310	Vapour pressure @21°C mmHg	123
Odour threshold (ppm)	0.1	POCP	

Aqueous effluent	
Solubility in water (25°C %w/w)	1.2
Solubility of water in (25°C %w/w)	0.62
Log_{10} activated carbon partition	2.9
Log_{10} partition in octanol/water (w/w)	+2.0
Biological oxygen demand w/w (days)	0.19 (5)
Theoretical oxygen demand w/w	2.83

Vapour pressure equation constants (Log_{10}, mmHg)		
Antoine equation	A	6.84953
	B	1139.34
	C	231.742
Cox chart	A	7.09624
	B	1256.2

Solvent properties			
Solubility parameter	6.9	Kauri butanol value	
Dipole (D)	1.2	Evaporation time (ether = 1)	1.6
Dielectric constant (20°C)		Evaporation time (BuAc = 1)	8.1
Polarity (water 100)			

Thermal information	
Latent heat (cal/mol)	6936
Nett heat of combustion (kcal/gmol)	885
Specific heat (cal/mol/°C)	52
Critical pressure (MN/m^2)	3.14
Critical temperature (K)	500
Latent heat of fusion (cal/mol)	2631
Van der Waals' volume	4.74
Van der Waals' surface area	4.09
Molar volume	142.3

Solute	Azeotrope X% w/w	°C	Reference	Solute γ^∞	Reference	Partition coefficient	Reference
Hydrocarbons							
n-Pentane							
n-Hexane	53	67	12128				
n-Heptane	None		12196	1.1	3+4/559		
n-Octane				1.3	1x/3/1243		
n-Nonane							
n-Decane							
2,2,4-TMP							
Cyclohexane				1.0	3+4/555		
Benzene	None		10863	1.2	3+4/553		
Toluene				1.1	3+4/558		
Ethylbenzene				1.4	3+4/563		
Xylenes							
C_9 Aromatics							
Tetralin							
Alcohols							
Methanol	76	57	2091a	3.3	2a/261		
Ethanol	83	64	4110	5.0	2a/459		
n-Propanol				3.5	2a/586		
i-Propanol	85	66	6391	4.5	2b/101	0.5	V2/618
n-Butanol				3.0	2b/202		
i-Butanol							
s-Butanol							
n-Amyl alc.							
i-Amyl alc.							
Cyclohexanol							
1-Octanol							
Ethanediol							
DEG							
1,2-Propanediol							
Glycol ethers							
PGME							
EGME							
EEE							
EGBE							
Chlorinated							
MDC							
Chloroform	64	71	1496	0.5	3+4/537		
Carbon tet.	None		1163	1.1	3+4/529		
1,2-EDC	None		3004				
1,1,1-TCA							
TCE	None		2331				
Perk.							
MCB							

Ethers

Solute	Azeotrope X% w/w	°C	Reference	Solute γ^∞	Reference	Partition coefficient	Reference
Ketones							
Acetone	39	54	5386				
MEK	Azeo		3b/357	2.8	3b/357		
MIBK							
Cyclohexanone				1.7	3b/506		
NMP							
Acetophenone							
Ethers							
Diethyl ether							
DIPE	–		–	–	–	–	–
Dibutyl ether							
MTBE							
1,4-Dioxane	None		7548	2.2	1x/3/1243		
THF							
Esters							
Me acetate							
Et acetate							
i-Propyl acetate							
n-Butyl acetate							
Cellosolve acetate							
Miscellaneous							
DMF							
DMAc						4.5	V3/103
DMSO							
Sulfolane							
CS_2							
Acetic acid	None		3185	3.2	3+4/544	0.6	V2/282
Aniline							
Nitrobenzene							
Morpholine							
Pyridine							
2-Nitropropane							
Acetonitrile							
Furfuraldehyde							
Phenol						0.01	P1656
Water	95	62	579	19.2	1/525		

Dibutyl ether

Alternative names
Butyl ether

Reference codes			
CAS number	142 96 1	Hazchem code	
UN number		EPA code	

Physical properties			
Molecular weight	130	Cubic expansion coeff (per °C × 10^3)	1.15
Empirical formula	$C_8H_{18}O_1$	Surface tension (@20°C dyn/cm)	1.4
Boiling point (°C)	142	Absolute viscosity (@25°C cP)	0.63
Freezing point (°C)	–95	Refractive index (25°C)	1.397
Specific gravity (20/4)	0.769		

Fire hazards			
Flash point (closed cup °C)	25	Lower explosive limit (ppm)	15000
Autoignition temperature (°C)	194	Upper explosive limit (ppm)	76000
Electrical conductivity			

Health hazards			
IDLH (ppm)		Vapour concentration @21°C ppm	7377
OES-TWA		Vapour density (relative to air)	4.48
OES-STEL		Vapour pressure @21°C mmHg	5.5
Odour threshold (ppm)	0.5	POCP	

Aqueous effluent	
Solubility in water (25°C %w/w)	0.03
Solubility of water in (25°C %w/w)	0.20
Log_{10} activated carbon partition	4.59
Log_{10} partition in octanol/water (w/w)	
Biological oxygen demand w/w (days)	
Theoretical oxygen demand w/w	2.95

Vapour pressure equation constants (Log_{10}, mmHg)		
Antoine equation	A	6.7963
	B	1297.29
	C	191.03
Cox chart	A	7.31357
	B	1649.0

Solvent properties			
Solubility parameter	7.2	Kauri butanol value	
Dipole (D)	1.2	Evaporation time (ether = 1)	
Dielectric constant (20°C)		Evaporation time (BuAc = 1)	
Polarity (water 100)	7.1		

Thermal information	
Latent heat (cal/mol)	8944
Nett heat of combustion (kcal/gmol)	1182
Specific heat (cal/mol/°C)	66.0
Critical pressure (MN/m^2)	
Critical temperature (K)	307
Latent heat of fusion (cal/mol)	
Van der Waals' volume	6.09
Van der Waals' surface area	5.18
Molar volume	170.4

Solute	Azeotrope X% w/w	°C	Reference	Solute γ^{∞}	Reference	Partition coefficient	Reference
Hydrocarbons							
n-Pentane				1.0	1x/3/1370		
n-Hexane				1.0	1x/3/1370		
n-Heptane				1.0	1x/3/1370		
n-Octane				1.1	1x/3/1370		
n-Nonane							
n-Decane							
2,2,4-TMP							
Cyclohexane				1.0	1x/3/1370		
Benzene				0.9	1x/3/1370		
Toluene				1.0	1x/3/1370		
Ethylbenzene	None		14102				
Xylenes	78	142	14117				
C_9 Aromatics							
Tetralin							
Alcohols							
Methanol							
Ethanol	None		4170	3.5	2e/391	0.98	V2/389
n-Propanol							
i-Propanol							
n-Butanol	17	118	8195	0.7	2d/231	0.04	V3/122
i-Butanol	None		8379				
s-Butanol							
n-Amyl alc.	50	135	9770				
i-Amyl alc.	35	130	9883				
Cyclohexanol	None		11745				
1-Octanol							
Ethanediol	90	140	4354				
DEG							
1,2-Propanediol		136	6668				
Glycol ethers							
PGME	63	138	10003				
EGME	32	122	6615				
EEE	50	127	8479				
EGBE	None		12247				
Chlorinated							
MDC							
Chloroform	None		1501a	0.5	1x/3/1370		
Carbon tet.				0.8	1x/3/1370		
1,2-EDC				1.2	1x/3/1370		
1,1,1-TCA				0.8	1x/3/1370		
TCE							
Perk.							
MCB	None		10537				

Ethers

Solute	Azeotrope X% w/w	Azeotrope °C	Azeotrope Reference	Solute γ^∞	Reference	Partition coefficient	Reference
Ketones							
Acetone						0.07	CEP
MEK				2.3	1x/3/1370		
MIBK							
Cyclohexanone							
NMP							
Acetophenone							
Ethers							
Diethyl ether	None		8305				
DIPE							
Dibutyl ether	–		–	–	–	–	–
MTBE							
1,4-Dioxane				2.1	1x/3/1370		
THF							
Esters							
Me acetate							
Et acetate							
i-Propyl acetate							
n-Butyl acetate	5	126	11333				
Cellosolve acetate							
Miscellaneous							
DMF							
DMAc						5.0	V3/104
DMSO							
Sulfolane							
CS_2							
Acetic acid	None		3220			1.0	V2/304
Aniline	None		11198				
Nitrobenzene							
Morpholine							
Pyridine							
2-Nitropropane							
Acetonitrile							
Furfuraldehyde	80	138	8788	4.2	3a/139		
Phenol	None		10960				
Water	67	93	735				

Methyl tert butyl ether

Alternative names
Tert butyl methyl ether, MTBE

Reference codes			
CAS number	1634 04 4	Hazchem code	
UN number		EPA code	

Physical properties			
Molecular weight	88	Cubic expansion coeff (per °C × 10^3)	
Empirical formula	$C_5H_{12}O_1$	Surface tension (@20°C dyn/cm)	18.3
Boiling point (°C)	55	Absolute viscosity (@25°C cP)	0.35
Freezing point (°C)	–109	Refractive index (20°C)	1.369
Specific gravity (20/4)	0.741		

Fire hazards			
Flash point (closed cup °C)	–34	Lower explosive limit (ppm)	16000
Autoignition temperature (°C)	460	Upper explosive limit (ppm)	84000
Electrical conductivity			

Health hazards			
IDLH (ppm)		Vapour concentration @21°C ppm	372000
OES-TWA		Vapour density (relative to air)	3.06
OES-STEL		Vapour pressure @21°C mmHg	206
Odour threshold (ppm)		POCP	

Aqueous effluent	
Solubility in water (25°C %w/w)	4.3
Solubility of water in (25°C %w/w)	1.4
Log_{10} activated carbon partition	
Log_{10} partition in octanol/water (w/w)	
Biological oxygen demand w/w (days)	
Theoretical oxygen demand w/w	2.75

Vapour pressure equation constants (Log_{10}, mmHg)		
Antoine equation	A	
	B	
	C	
Cox chart	A	7.06046
	B	1191.2

Solvent properties			
Solubility parameter	7.4	Kauri butanol value	
Dipole (D)	1.2	Evaporation time (ether = 1)	1.6
Dielectric constant (20°C)	4.5	Evaporation time (BuAc = 1)	8.4
Polarity (water 100)	14.8		

Thermal information	
Latent heat (cal/mol)	7030
Nett heat of combustion (kcal/gmol)	740
Specific heat (cal/mol/°C)	
Critical pressure (MN/m^2)	
Critical temperature (K)	
Latent heat of fusion (cal/mol)	
Van der Waals' volume	4.07
Van der Waals' surface area	3.63
Molar volume	119.0

Solute	Azeotrope X% w/w	Azeotrope °C	Reference	Solute γ^{∞}	Reference	Partition coefficient	Reference
Hydrocarbons							
n-Pentane							
n-Hexane				1.2	1x/3/1152		
n-Heptane				0.9	1x/3/1152		
n-Octane							
n-Nonane							
n-Decane							
2,2,4-TMP							
Cyclohexane				1.1	1x/3/1152		
Benzene							
Toluene							
Ethylbenzene							
Xylenes							
C_9 Aromatics							
Tetralin							
Alcohols							
Methanol	90	51	2058	3.0	2c/160		
Ethanol				3.0			
n-Propanol				2.9			
i-Propanol				2.6			
n-Butanol							
i-Butanol							
s-Butanol							
n-Amyl alc.							
i-Amyl alc.							
Cyclohexanol							
1-Octanol							
Ethanediol							
DEG							
1,2-Propanediol							
Glycol ethers							
PGME							
EGME							
EEE							
EGBE							
Chlorinated							
MDC							
Chloroform							
Carbon tet.							
1,2-EDC							
1,1,1-TCA							
TCE							
Perk.							
MCB							

Ethers

Solute	Azeotrope X% w/w	°C	Reference	Solute γ^∞	Reference	Partition coefficient	Reference
Ketones							
Acetone							
MEK							
MIBK							
Cyclohexanone							
NMP							
Acetophenone							
Ethers							
Diethyl ether							
DIPE							
Dibutyl ether							
MTBE	–		–	–	–	–	–
1,4-Dioxane							
THF							
Esters							
Me acetate							
Et acetate							
i-Propyl acetate							
n-Butyl acetate							
Cellosolve acetate							
Miscellaneous							
DMF							
DMAc							
DMSO							
Sulfolane							
CS_2							
Acetic acid							
Aniline							
Nitrobenzene							
Morpholine							
Pyridine							
2-Nitropropane							
Acetonitrile							
Furfuraldehyde							
Phenol							
Water	97	52	466				

1,4-Dioxane

Alternative names
Glycol ethylene ether, p-dioxane, diethylene dioxide, diethylene oxide

Reference codes

CAS number	123 91 1	Hazchem code	2SE
UN number	1185	EPA code	

Physical properties

Molecular weight	88	Cubic expansion coeff (per °C × 10^3)	1.1
Empirical formula	$C_4H_8O_2$	Surface tension (@20°C dyn/cm)	40
Boiling point (°C)	101	Absolute viscosity (@25°C cP)	1.3
Freezing point (°C)	+12	Refractive index (25°C)	1.420
Specific gravity (20/4)	1.034		

Fire hazards

Flash point (closed cup °C)	12	Lower explosive limit (ppm)	20000
Autoignition temperature (°C)	180	Upper explosive limit (ppm)	222000
Electrical conductivity	5E–15		

Health hazards

IDLH (ppm)	200	Vapour concentration @21°C ppm	41000
OES-TWA	25	Vapour density (relative to air)	3.06
OES-STEL	100	Vapour pressure @21°C mmHg	32
Odour threshold (ppm)	170	POCP	

Aqueous effluent

Solubility in water (25°C %w/w)	Total
Solubility of water in (25°C %w/w)	Total
Log_{10} activated carbon partition	
Log_{10} partition in octanol/water (w/w)	–0.42
Biological oxygen demand w/w (days)	0
Theoretical oxygen demand w/w	1.82

Vapour pressure equation constants (Log_{10}, mmHg)

Antoine equation	A	7.43155
	B	1554.679
	C	240.337
Cox chart	A	7.19047
	B	1426.5

Solvent properties

Solubility parameter	10.0	Kauri butanol value	
Dipole (D)	0.4	Evaporation time (ether = 1)	7.3
Dielectric constant (20°C)	2.21	Evaporation time (BuAc = 1)	2.2
Polarity (water 100)	16.4		

Thermal information

Latent heat (cal/mol)	8510
Nett heat of combustion (kcal/gmol)	567
Specific heat (cal/mol/°C)	36
Critical pressure (MN/m^2)	5.21
Critical temperature (K)	588
Latent heat of fusion (cal/mol)	3080
Van der Waals' volume	3.19
Van der Waals' surface area	2.64
Molar volume	85.1

Solute	Azeotrope X% w/w	Azeotrope °C	Azeotrope Reference	Solute γ^{∞}	Reference	Partition coefficient	Reference
Hydrocarbons							
n-Pentane							
n-Hexane	2	60	7547	3.3	3+4/472		
n-Heptane	44	92	7552	4.0	1x/3/1052		
n-Octane		100	7554	7.4	1x/3/1052		
n-Nonane				4.7	3+4/481		
n-Decane							
2,2,4-TMP							
Cyclohexane	25	80	7540	2.6	3+4/468		
Benzene	12	82	7537	1.1	3+4/465		
Toluene	80	102	7550	1.2	1x/3/1052		
Ethylbenzene							
Xylenes							
C_9 Aromatics							
Tetralin							
Alcohols							
Methanol	None		1998	1.6	2a/148		
Ethanol	9	78	4011	2.5	1x/3/1052		
n-Propanol	45	95	6447	1.7	2a/531		
i-Propanol	None		6337	1.2	2b/56		
n-Butanol	None		7519	1.5	2b/147		
i-Butanol	96	101	7522	1.4	2b/278		
s-Butanol	60	99	7520				
n-Amyl alc.							
i-Amyl alc.							
Cyclohexanol	None		7543				
1-Octanol							
Ethanediol	None		4206				
DEG							
1,2-Propanediol							
Glycol ethers							
PGME							
EGME	None		6541				
EEE							
EGBE							
Chlorinated							
MDC							
Chloroform	None		1465	0.46	3+4/441		
Carbon tet.	None		1124	1.2	1x/3/1052		
1,2-EDC	None		2979	1.4	3+4/447		
1,1,1-TCA							
TCE	None		2301				
Perk.	None		2181				
MCB							

Ethers

Solute	Azeotrope X% w/w	°C	Reference	Solute γ^∞	Reference	Partition coefficient	Reference
Ketones							
Acetone	None		5333				
MEK				1.3	1x/3/1052		
MIBK				1.6	3b/523		
Cyclohexanone	None		7539				
NMP							
Acetophenone							
Ethers							
Diethyl ether							
DIPE	None		7548				
Dibutyl ether							
MTBE							
1,4-Dioxane	–		–	–	–	–	–
THF							
Esters							
Me acetate							
Et acetate	None		7514	1.1	3+4/455		
i-Propyl acetate	None		7527				
n-Butyl acetate							
Cellosolve acetate							
Miscellaneous							
DMF				1.7	3+4/454		
DMAc							
DMSO				2.2	3+4/450		
Sulfolane							
CS_2				2.4	3+4/446		
Acetic acid	23	119	119	2.3	3+4/448		
Aniline							
Nitrobenzene							
Morpholine							
Pyridine	None		7524				
2-Nitropropane							
Acetonitrile	None		2769a				
Furfuraldehyde							
Phenol							
Water	82	88	349	4.1	1/382		

Tetrahydrofuran

Alternative names
THF, tetramethylene oxide, 1,4-epoxybutane, oxacyclopentane

Reference codes

CAS number	109 99 9	Hazchem code	2SE
UN number	2056	EPA code	U213

Physical properties

Molecular weight	72	Cubic expansion coeff (per °C × 10^3)	1.1
Empirical formula	$C_4H_8O_1$	Surface tension (@20°C dyn/cm)	28
Boiling point (°C)	66	Absolute viscosity (@25°C cP)	0.55
Freezing point (°C)	–109	Refractive index (25°C)	1.404
Specific gravity (20/4)	0.888		

Fire hazards

Flash point (closed cup °C)	–15	Lower explosive limit (ppm)	23000
Autoignition temperature (°C)	212	Upper explosive limit (ppm)	118000
Electrical conductivity	4.5 E–5		

Health hazards

IDLH (ppm)		Vapour concentration @21°C ppm	230000
OES-TWA	100	Vapour density (relative to air)	2.5
OES-STEL	200	Vapour pressure @21°C mmHg	133
Odour threshold (ppm)	30	POCP	70

Aqueous effluent

Solubility in water (25°C %w/w)	Total	Lower critical solution temperature 72°C
Solubility of water in (25°C %w/w)	Total	
Log_{10} activated carbon partition		
Log_{10} partition in octanol/water (w/w)	+0.46	
Biological oxygen demand w/w (days)		
Theoretical oxygen demand w/w	2.59	

Vapour pressure equation constants (Log_{10}, mmHg)

Antoine equation	A	6.99515
	B	1202.29
	C	226.254
Cox chart	A	7.09092
	B	1246.2

Solvent properties

Solubility parameter	9.1	Kauri butanol value	
Dipole (D)	1.75	Evaporation time (ether = 1)	2.3
Dielectric constant (20°C)	7.6	Evaporation time (BuAc = 1)	6.3
Polarity (water 100)	21		

Thermal information

Latent heat (cal/mol)	6664
Nett heat of combustion (kcal/gmol)	601
Specific heat (cal/mol/°C)	36
Critical pressure (MN/m^2)	5.2
Critical temperature (K)	541
Latent heat of fusion (cal/mol)	
Van der Waals' volume	2.94
Van der Waals' surface area	2.72
Molar volume	81.08

Solute	Azeotrope X% w/w	Azeotrope °C	Azeotrope Reference	Solute γ^{∞}	Reference	Partition coefficient	Reference
Hydrocarbons							
n-Pentane	None			2.0	1x/3/1046		
n-Hexane	50	63	7407	1.9	1x/3/1046		
n-Heptane							
n-Octane				1.7	1x/3/1047		
n-Nonane							
n-Decane							
2,2,4-TMP							
Cyclohexane	97	60		1.7	1x/3/1046		
Benzene				0.84	1x/3/1046		
Toluene				0.85	1x/3/1046		
Ethylbenzene				0.90	1x/3/1047		
Xylenes							
C_9 Aromatics				0.9	1x/3/1047		
Tetralin							
Alcohols							
Methanol	69	61	1996	2.4	2a/141		
Ethanol	90	66	4009	1.9	2c/328		
n-Propanol	None			1.4	2c/497		
i-Propanol	None		6335a	1.4	2b/55		
n-Butanol	None			1.2	2b/146		
i-Butanol							
s-Butanol							
n-Amyl alc.							
i-Amyl alc.							
Cyclohexanol							
1-Octanol							
Ethanediol	None		4204a	6.6	2d/3		
DEG							
1,2-Propanediol							
Glycol ethers							
PGME							
EGME							
EEE							
EGBE							
Chlorinated							
MDC				0.5	1x/1/100		
Chloroform	34	72	1464	0.25	1x/3/1046		
Carbon tet.	None			0.75	3+4/429		
1,2-EDC				0.60	1x/3/1046		
1,1,1-TCA							
TCE							
Perk.							
MCB							

Ethers

Solute	Azeotrope X% w/w	Azeotrope °C	Azeotrope Reference	Solute γ^{∞}	Reference	Partition coefficient	Reference
Ketones							
Acetone	8	64					
MEK							
MIBK							
Cyclohexanone							
NMP							
Acetophenone							
Ethers							
Diethyl ether							
DIPE							
Dibutyl ether							
MTBE							
1,4-Dioxane				1.1	1x/3/1046		
THF	–		–	–	–	–	–
Esters							
Me acetate							
Et acetate				1.1	1x/3/1046		
i-Propyl acetate							
n-Butyl acetate							
Cellosolve acetate							
Miscellaneous							
DMF							
DMAc							
DMSO	None			4.6	3+4/433	2.05	V2/396
Sulfolane							
CS_2							
Acetic acid							
Aniline							
Nitrobenzene							
Morpholine							
Pyridine							
2-Nitropropane	None		6273				
Acetonitrile							
Furfuraldehyde							
Phenol							
Water	96	64	345	10.4	1/367		

Section 7

Esters

Methyl acetate

Alternative names
Acetic acid, methyl ester

Reference codes

CAS number	79 20 9	Hazchem code	2SE
UN number	1231	EPA code	

Physical properties

Molecular weight	74	Cubic expansion coeff (per °C × 10^3)	1.4
Empirical formula	$C_3H_6O_2$	Surface tension (@20°C dyn/cm)	24
Boiling point (°C)	57	Absolute viscosity (@25°C cP)	0.37
Freezing point (°C)	–98	Refractive index (25°C)	1.360
Specific gravity (20/4)	0.927		

Fire hazards

Flash point (closed cup °C)	–10	Lower explosive limit (ppm)	31000
Autoignition temperature (°C)	500	Upper explosive limit (ppm)	160000
Electrical conductivity	3.4E–6		

Health hazards

IDLH (ppm)	10000	Vapour concentration @21°C ppm	290000
OES-TWA	200	Vapour density (relative to air)	2.57
OES-STEL	250	Vapour pressure @21°C mmHg	171
Odour threshold (ppm)	200	POCP	2.5

Aqueous effluent

Solubility in water (25°C %w/w)	24.5
Solubility of water in (25°C %w/w)	8.2
Log_{10} activated carbon partition	1.85
Log_{10} partition in octanol/water (w/w)	+0.18
Biological oxygen demand w/w (days)	
Theoretical oxygen demand w/w	1.51

Vapour pressure equation constants (Log_{10}, mmHg)

Antoine equation	A	7.06524
	B	1157.63
	C	219.726
Cox chart	A	7.25014
	B	1254.0

Solvent properties

Solubility parameter	9.6	Kauri butanol value	
Dipole (D)	1.7	Evaporation time (ether = 1)	2.1
Dielectric constant (20°C)	6.7	Evaporation time (BuAc = 1)	9.5
Polarity (water 100)	29		

Thermal information

Latent heat (cal/mol)	7178
Nett heat of combustion (kcal/gmol)	348
Specific heat (cal/mol/°C)	37
Critical pressure (MN/m^2)	4.6
Critical temperature (K)	507
Latent heat of fusion (cal/mol)	
Van der Waals' volume	2.80
Van der Waals' surface area	2.58
Molar volume	79.8

Solute	Azeotrope X% w/w	°C	Reference	Solute γ^{∞}	Reference	Partition coefficient	Reference
Hydrocarbons							
n-Pentane	22	34	5536				
n-Hexane	63	52	5554				
n-Heptane	96	57	5558	5.5	1x/3/994		
n-Octane							
n-Nonane							
n-Decane							
2,2,4-TMP							
Cyclohexane	80	55	5541	3.6	5/393		
Benzene	0.3	57	5537	1.4	5/375		
Toluene							
Ethylbenzene							
Xylenes							
C_9 Aromatics							
Tetralin							
Alcohols							
Methanol	81	54	1967	2.7	2a/92		
Ethanol	97	57	3969	1.9	2a/335		
n-Propanol	None			2.8	2a/530		
i-Propanol	None		5516	2.2	2b/50		
n-Butanol				2.3	2f/137		
i-Butanol							
s-Butanol							
n-Amyl alc.							
i-Amyl alc.							
Cyclohexanol							
1-Octanol							
Ethanediol						3.81	V2/416
DEG							
1,2-Propanediol							
Glycol ethers							
PGME							
EGME							
EEE							
EGBE							
Chlorinated							
MDC	None		1557	0.6	5/347		
Chloroform	23	65	1448	0.6	5/341		
Carbon tet.	None		1110	1.7	5/339		
1,2-EDC							
1,1,1-TCA							
TCE							
Perk.							
MCB				1.2	5/374		

Esters

Solute	Azeotrope X% w/w	Azeotrope °C	Azeotrope Reference	Solute γ^∞	Reference	Partition coefficient	Reference
Ketones							
Acetone	50	55	5310	1.3	3+4/159	0.37	V2/463
MEK	None		5519	1.0	3+4/271		
MIBK							
Cyclohexanone							
NMP							
Acetophenone							
Ethers							
Diethyl ether	None		5527				
DIPE							
Dibutyl ether							
MTBE							
1,4-Dioxane							
THF							
Esters							
Me acetate	–		–	–	–	–	–
Et acetate	None		5521a	0.8	5/357		
i-Propyl acetate							
n-Butyl acetate	None			1.2	5/397		
Cellosolve acetate							
Miscellaneous							
DMF							
DMAc							
DMSO							
Sulfolane							
CS_2	30	40	1198	2.7	5/349		
Acetic acid	None		3101a	1.1	5/82	0.41	V4/183
Aniline							
Nitrobenzene							
Morpholine							
Pyridine							
2-Nitropropane							
Acetonitrile	None		2763	1.4	5/354		
Furfuraldehyde							
Phenol							
Water	97	56	276	8.5	1/264		

Ethyl acetate

Alternative names
ETAC, acetic ester, EtOAc, ethyl ethanoate, acetic acid, ethyl ester

Reference codes

CAS number	141 78 6	Hazchem code	3YE
UN number	1173	EPA code	U112

Physical properties

Molecular weight	88	Cubic expansion coeff (per °C × 10^3)	1.39
Empirical formula	$C_4H_8O_2$	Surface tension (@20°C dyn/cm)	24
Boiling point (°C)	77	Absolute viscosity (@25°C cP)	0.46
Freezing point (°C)	–84	Refractive index (25°C)	1.370
Specific gravity (20/4)	0.895		

Fire hazards

Flash point (closed cup °C)	–4	Lower explosive limit (ppm)	22000
Autoignition temperature (°C)	484	Upper explosive limit (ppm)	115000
Electrical conductivity	1.0E–9		

Health hazards

IDLH (ppm)	10000	Vapour concentration @21°C ppm	114000
OES-TWA	400	Vapour density (relative to air)	3.04
OES-STEL		Vapour pressure @21°C mmHg	78
Odour threshold (ppm)	50	POCP	21.8

Aqueous effluent

Solubility in water (25°C %w/w)	7.7
Solubility of water in (25°C %w/w)	3.3
Log_{10} activated carbon partition	2.31
Log_{10} partition in octanol/water (w/w)	+0.73
Biological oxygen demand w/w (days)	1.2
Theoretical oxygen demand w/w	1.82

Vapour pressure equation constants (Log_{10}, mmHg)

Antoine equation	A	7.10179
	B	1244.95
	C	217.881
Cox chart	A	7.30648
	B	1358.7

Solvent properties

Solubility parameter	9.1	Kauri butanol value	
Dipole (D)	1.7	Evaporation time (ether = 1)	3.0
Dielectric constant (20°C)	6.02	Evaporation time (BuAc = 1)	4.2
Polarity (water 100)	23		

Thermal information

Latent heat (cal/mol)	7744
Nett heat of combustion (kcal/gmol)	493
Specific heat (cal/mol/°C)	40
Critical pressure (MN/m^2)	3.84
Critical temperature (K)	523
Latent heat of fusion (cal/mol)	2494
Van der Waals' volume	3.48
Van der Waals' surface area	3.12
Molar volume	98.5

Solute	Azeotrope X% w/w	°C	Reference	Solute γ^{∞}	Reference	Partition coefficient	Reference
Hydrocarbons							
n-Pentane				3.1	1x/1/105		
n-Hexane	38	65	7588	2.4	5/514		
n-Heptane				2.9	1x/3/1051		
n-Octane				3.1	1x/3/1051		
n-Nonane							
n-Decane							
2,2,4-TMP				1.5	1x/3/1051		
Cyclohexane	54	72	7583	2.3	5/506		
Benzene	None		7580	3.1	5/502		
Toluene	None		7591	1.2	5/516		
Ethylbenzene				1.8	5/540		
Xylenes	None		7594	1.6	5/541		
C_9 Aromatics							
Tetralin							
Alcohols							
Methanol	56	62	1999	2.7	1x/1/103	1.08	V2/93
Ethanol	69	72	4012	2.2	2a/351	0.41	CEH
n-Propanol	None		6448	1.9	2a/536	0.19	V2/549
i-Propanol	75	76	6338	1.6	2b/59	0.17	CEH
n-Butanol	None		7567	2.3	2b/148	0.03	V3/50
i-Butanol	None		7570			0.04	V3/54
s-Butanol	None		7568			0.07	V3/52
n-Amyl alc.							
i-Amyl alc.							
Cyclohexanol	None			3.5	2d/511		
1-Octanol							
Ethanediol							
DEG							
1,2-Propanediol							
Glycol ethers							
PGME							
EGME				2.0	2b/126		
EEE	None		7571	2.2	2f/335		
EGBE	None		7590				
Chlorinated							
MDC				0.5	1x/1/103		
Chloroform	72	78	1466	0.5	1x/1/103		
Carbon tet.	43	75	1125	1.3	1x/1/103		
1,2-EDC	None		2980	0.8	1x/1/103		
1,1,1-TCA							
TCE	None		2302	0.9	5/454		
Perk.							
MCB	None		7578	1.4	5/492		

Esters

Solute	Azeotrope X% w/w	Azeotrope °C	Azeotrope Reference	Solute γ^{∞}	Reference	Partition coefficient	Reference
Ketones							
Acetone	None		5334	1.2	3+4/176	0.14	CEH
MEK	82	77	7345	2.0	3+4/278		
MIBK				0.6	3b/527		
Cyclohexanone							
NMP							
Acetophenone							
Ethers							
Diethyl ether				1.0	3+4/513		
DIPE							
Dibutyl ether							
MTBE							
1,4-Dioxane	None		7514	1.1	3+4/455		
THF				1.1	1x/1/104		
Esters							
Me acetate	None		5521a	1.1	5/357		
Et acetate	–		–	–	–	–	–
i-Propyl acetate	None			1.0	5/487		
n-Butyl acetate	None		7585				
Cellosolve acetate							
Miscellaneous							
DMF							
DMAc						1.85	V3/48
DMSO				3.9	5/461		
Sulfolane							
CS_2	3	46	1220	2.7	1x/1/103		
Acetic acid	None		3120	7.0	5/104	0.29	V4/184
Aniline							
Nitrobenzene							
Morpholine							
Pyridine							
2-Nitropropane	None		6274				
Acetonitrile	77	75	2770	1.6	1x/1/103	0.17	V2/175
Furfuraldehyde	None		7574	1.7	3a/123	0.03	V2/57
Phenol						<0.01	V4/241
Water	91.5	70	352	9.7	1/393		

Isopropyl acetate

Alternative names
Sec propyl acetate, 2-propyl acetate, acetic acid, isopropyl ester

Reference codes

CAS number	108 21 4	Hazchem code	3YE
UN number	1220	EPA code	

Physical properties

Molecular weight	102	Cubic expansion coeff (per °C × 10^3)	1.31
Empirical formula	$C_5H_{10}O_2$	Surface tension (@20°C dyn/cm)	22.1
Boiling point (°C)	89	Absolute viscosity (@25°C cP)	0.46
Freezing point (°C)	–69	Refractive index (25°C)	1.375
Specific gravity (20/4)	0.874		

Fire hazards

Flash point (closed cup °C)	3	Lower explosive limit (ppm)	18000
Autoignition temperature (°C)	460	Upper explosive limit (ppm)	80000
Electrical conductivity	5.7E–7		

Health hazards

IDLH (ppm)	16000	Vapour concentration @21°C ppm	66000
OES-TWA	250	Vapour density (relative to air)	3.5
OES-STEL	200	Vapour pressure @21°C mmHg	47
Odour threshold (ppm)	30	POCP	21.5

Aqueous effluent

Solubility in water (25°C %w/w)	2.9
Solubility of water in (25°C %w/w)	3.2
Log_{10} activated carbon partition	2.63
Log_{10} partition in octanol/water (w/w)	
Biological oxygen demand w/w (days)	
Theoretical oxygen demand w/w	2.04

Vapour pressure equation constants (Log_{10}, mmHg)

Antoine equation	A	7.3340
	B	1436.53
	C	233.7
Cox chart	A	7.34068
	B	1422.7

Solvent properties

Solubility parameter	8.5	Kauri butanol value	
Dipole (D)	2.7	Evaporation time (ether = 1)	
Dielectric constant (20°C)		Evaporation time (BuAc = 1)	
Polarity (water 100)			

Thermal information

Latent heat (cal/mol)	8262
Nett heat of combustion (kcal/gmol)	534
Specific heat (cal/mol/°C)	50
Critical pressure (MN/m^2)	3.65
Critical temperature (K)	538
Latent heat of fusion (cal/mol)	
Van der Waals' volume	4.15
Van der Waals' surface area	3.65
Molar volume	117.8

Solute	Azeotrope X% w/w	°C	Reference	Solute γ^∞	Reference	Partition coefficient	Reference
Hydrocarbons							
n-Pentane							
n-Hexane	9	69	9297				
n-Heptane	67	88	9302				
n-Octane							
n-Nonane							
n-Decane							
2,2,4-TMP							
Cyclohexane	25	79	9296				
Benzene	None		9294				
Toluene	None		9300				
Ethylbenzene							
Xylenes							
C_9 Aromatics							
Tetralin							
Alcohols							
Methanol	20	65	2046				
Ethanol	48	77	4054	1.9	2a/391		
n-Propanol							
i-Propanol	48	80	6363	1.7	2f/59		
n-Butanol	None		8121				
i-Butanol							
s-Butanol							
n-Amyl alc.							
i-Amyl alc.							
Cyclohexanol							
1-Octanol							
Ethanediol							
DEG							
1,2-Propanediol							
Glycol ethers							
PGME							
EGME							
EEE							
EGBE							
Chlorinated							
MDC							
Chloroform							
Carbon tet.	None		1145				
1,2-EDC	None		2992				
1,1,1-TCA							
TCE	None		2317				
Perk.							
MCB							

Esters

Solute	Azeotrope X% w/w	°C	Reference	Solute γ^{∞}	Reference	Partition coefficient	Reference
Ketones							
Acetone	None		5365				
MEK	None		7366				
MIBK							
Cyclohexanone							
NMP							
Acetophenone							
Ethers							
Diethyl ether							
DIPE							
Dibutyl ether							
MTBE							
1,4-Dioxane	None		7527				
THF							
Esters							
Me acetate							
Et acetate	None			1.0	5/487		
i-Propyl acetate	–		–	–	–	–	–
n-Butyl acetate							
Cellosolve acetate							
Miscellaneous							
DMF							
DMAc							
DMSO							
Sulfolane							
CS_2	None		1251				
Acetic acid	None		3147	1.4	5/123	0.36	V2/236
Aniline							
Nitrobenzene							
Morpholine							
Pyridine							
2-Nitropropane							
Acetonitrile	20	80	2788				
Furfuraldehyde							
Phenol						<0.01	V4/264
Water	90	77	439				

n-Butyl acetate

Alternative names
BuAc, *n*-butyl ethanoate, BuOAc, acetic acid, butyl ester

Reference codes

CAS number	123 86 4	Hazchem code	3YE
UN number	1123	EPA code	

Physical properties

Molecular weight	116	Cubic expansion coeff (per °C × 10^3)	1.16
Empirical formula	$C_6H_{12}O_2$	Surface tension (@20°C dyn/cm)	25.1
Boiling point (°C)	126	Absolute viscosity (@25°C cP)	0.73
Freezing point (°C)	–73	Refractive index (25°C)	1.392
Specific gravity (20/4)	0.876		

Fire hazards

Flash point (closed cup °C)	22	Lower explosive limit (ppm)	17000
Autoignition temperature (°C)	407	Upper explosive limit (ppm)	150000
Electrical conductivity	1.6E–8		

Health hazards

IDLH (ppm)	10000	Vapour concentration @21°C ppm	14200
OES-TWA	150	Vapour density (relative to air)	4.03
OES-STEL	200	Vapour pressure @21°C mmHg	10.6
Odour threshold (ppm)	15	POCP	32.3

Aqueous effluent

Solubility in water (25°C %w/w)	0.7
Solubility of water in (25°C %w/w)	1.3
Log_{10} activated carbon partition	3.04
Log_{10} partition in octanol/water (w/w)	+1.7
Biological oxygen demand w/w (days)	1.15
Theoretical oxygen demand w/w	2.21

Vapour pressure equation constants (Log_{10}, mmHg)

Antoine equation	A	7.02845
	B	1368.50
	C	204.00
Cox chart	A	7.44951
	B	1626.5

Solvent properties

Solubility parameter	8.6	Kauri butanol value	
Dipole (D)	1.8	Evaporation time (ether = 1)	11.8
Dielectric constant (20°C)		Evaporation time (BuAc = 1)	1.0
Polarity (water 100)	24.1		

Thermal information

Latent heat (cal/mol)	8584
Nett heat of combustion (kcal/gmol)	784
Specific heat (cal/mol/°C)	58
Critical pressure (MN/m^2)	3.05
Critical temperature (K)	579
Latent heat of fusion (cal/mol)	
Van der Waals' volume	4.83
Van der Waals' surface area	4.20
Molar volume	132.5

Solute	Azeotrope X% w/w	Azeotrope °C	Azeotrope Reference	Solute γ^∞	Reference	Partition coefficient	Reference
Hydrocarbons							
n-Pentane							
n-Hexane							
n-Heptane	None		11826a	1.8	5/591		
n-Octane	52	119	11832				
n-Nonane							
n-Decane							
2,2,4-TMP							
Cyclohexane	None		11687	1.5	5/585		
Benzene	None		10859	0.9	5/583		
Toluene				1.0	5/586		
Ethylbenzene	None		11828				
Xylenes	None		11830				
C_9 Aromatics							
Tetralin							
Alcohols							
Methanol	None			5.8	2c/213	1.12	V2/131
Ethanol	None			2.1	2c/426	0.61	V2/363
n-Propanol	60	94		1.2	2e/484	0.11	V2/574
i-Propanol	None			1.8	2d/75		
n-Butanol	37	116	8153	1.4	2b/197	0.03	V4/253
i-Butanol	None		8345				
s-Butanol	None		8237				
n-Amyl alc.	None		9754				
i-Amyl alc.	82	126	9837				
Cyclohexanol	None			1.5	2f/417		
1-Octanol							
Ethanediol	None		4258	6.9	2d/15		
DEG							
1,2-Propanediol	None		6655a				
Glycol ethers							
PGME	None		9968				
EGME	52	119	6576	2.3	2d/122		
EEE	87	126	8434	1.8	2b/294		
EGBE	None		11823				
Chlorinated							
MDC							
Chloroform	None		1493	0.7	5/574		
Carbon tet.	None		1161	1.2	5/573		
1,2-EDC							
1,1,1-TCA							
TCE				0.7	5/575		
Perk.	21	120	2210				
MCB	None		10516				

Esters

Solute	Azeotrope X% w/w	Azeotrope °C	Azeotrope Reference	Solute γ^{∞}	Reference	Part coefficient	Reference
Ketones							
Acetone	None		5383	1.4	3b/197	0.14	CEH
MEK							
MIBK							
Cyclohexanone							
NMP							
Acetophenone							
Ethers							
Diethyl ether							
DIPE							
Dibutyl ether	95	126	11833				
MTBE							
1,4-Dioxane							
THF							
Esters							
Me acetate	None			1.2	5/397		
Et acetate	None		7585				
i-Propyl acetate							
n-Butyl acetate	–		–	–	–	–	–
Cellosolve acetate	None		11821				
Miscellaneous							
DMF							
DMAc							
DMSO							
Sulfolane							
CS_2							
Acetic acid	None		3177	2.1	5/147	0.23	V4/185
Aniline							
Nitrobenzene							
Morpholine							
Pyridine	None		8850				
2-Nitropropane							
Acetonitrile				2.3	5/577	0.07	V2/184
Furfuraldehyde				1.8	3+4/46	0.02	V3/194
Phenol	Azeo		10896	0.5	2b/373	<0.01	V3/297
Water	71	90	542	6.7	1/516		

Cellosolve acetate

Alternative names
Ethylene glycol monoethyl acetate, 2-ethoxy ethyl acetate

Reference codes

CAS number	111 15 9	Hazchem code	2 (S)
UN number	1172	EPA code	

Physical properties

Molecular weight	132	Cubic expansion coeff (per °C × 10^3)	1.12
Empirical formula	$C_6H_{12}O_3$	Surface tension (@20°C dyn/cm)	28.2
Boiling point (°C)	156	Absolute viscosity (@25°C cP)	
Freezing point (°C)	–62	Refractive index (20°C)	1.403
Specific gravity (20/4)	0.973		

Fire hazards

Flash point (closed cup °C)	52	Lower explosive limit (ppm)	17000
Autoignition temperature (°C)	345	Upper explosive limit (ppm)	82000
Electrical conductivity			

Health hazards

IDLH (ppm)	2500	Vapour concentration @21°C ppm	2108
OES-TWA	5	Vapour density (relative to air)	4.6
OES-STEL		Vapour pressure @21°C mmHg	1.5
Odour threshold (ppm)	0.1	POCP	60

Aqueous effluent

Solubility in water (25°C %w/w)	22.9
Solubility of water in (25°C %w/w)	6.5
Log_{10} activated carbon partition	
Log_{10} partition in octanol/water (w/w)	
Biological oxygen demand w/w (days)	
Theoretical oxygen demand w/w	1.82

Vapour pressure equation constants (Log_{10}, mmHg)

Antoine equation	A	6.8246
	B	1291.3
	C	170.97
Cox chart	A	7.542
	B	1799.3

Solvent properties

Solubility parameter	8.6	Kauri butanol value	
Dipole (D)	1.8	Evaporation time (ether = 1)	57
Dielectric constant (20°C)		Evaporation time (BuAc = 1)	0.2
Polarity (water 100)			

Thermal information

Latent heat (cal/mol)	9768
Nett heat of combustion (kcal/gmol)	754
Specific heat (cal/mol/°C)	63.9
Critical pressure (MN/m^2)	3.0
Critical temperature (K)	607
Latent heat of fusion (cal/mol)	
Van der Waals' volume	
Van der Waals' surface area	
Molar volume	135.7

Solute	Azeotrope X% w/w	°C	Reference	Solute γ^∞	Reference	Partition coefficient	Reference
Hydrocarbons							
n-Pentane							
n-Hexane							
n-Heptane							
n-Octane							
n-Nonane							
n-Decane							
2,2,4-TMP							
Cyclohexane							
Benzene							
Toluene	None		11913				
Ethylbenzene	None		12007				
Xylenes	None		12009				
C_9 Aromatics							
Tetralin							
Alcohols							
Methanol							
Ethanol							
n-Propanol							
i-Propanol							
n-Butanol	None		8160				
i-Butanol	None		8352				
s-Butanol							
n-Amyl alc.							
i-Amyl alc.	None		9842				
Cyclohexanol							
1-Octanol							
Ethanediol	None		4261				
DEG							
1,2-Propanediol	None		6665				
Glycol ethers							
PGME	13	151	9969				
EGME	None		6583				
EEE	None		8440				
EGBE	None		11989				
Chlorinated							
MDC							
Chloroform							
Carbon tet.							
1,2-EDC							
1,1,1-TCA							
TCE							
Perk.							
MCB							

Esters

Solute	Azeotrope X% w/w	°C	Reference	Solute γ^∞	Reference	Partition coefficient	Reference
Ketones							
Acetone							
MEK							
MIBK							
Cyclohexanone							
NMP							
Acetophenone							
Ethers							
Diethyl ether							
DIPE							
Dibutyl ether	88	142	12021				
MTBE							
1,4-Dioxane							
THF							
Esters							
Me acetate							
Et acetate							
i-Propyl acetate							
n-Butyl acetate	None		11821				
Cellosolve acetate	–		–	–	–	–	–
Miscellaneous							
DMF							
DMAc							
DMSO							
Sulfolane							
CS_2							
Acetic acid							
Aniline							
Nitrobenzene							
Morpholine							
Pyridine							
2-Nitropropane							
Acetonitrile							
Furfuraldehyde	None		8766				
Phenol	28	185	10898				
Water	50	97	559				

Section 8

Miscellaneous solvents

Dimethylformamide

Alternative names
DMF

Reference codes

CAS number	68 12 2	Hazchem code	2P
UN number	2265	EPA code	

Physical properties

Molecular weight	73	Cubic expansion coeff (per °C × 10^3)	1.03
Empirical formula	$C_3H_7N_1O_1$	Surface tension (@20°C dyn/cm)	35
Boiling point (°C)	153	Absolute viscosity (@25°C cP)	0.82
Freezing point (°C)	–61	Refractive index (25°C)	1.427
Specific gravity (20/4)	0.945		

Fire hazards

Flash point (closed cup °C)	62	Lower explosive limit (ppm)	22000
Autoignition temperature (°C)	445	Upper explosive limit (ppm)	160000
Electrical conductivity	6.0E–8		

Health hazards

IDLH (ppm)	3500	Vapour concentration @21°C ppm	3700
OES-TWA	10	Vapour density (relative to air)	2.53
OES-STEL	20	Vapour pressure @21°C mmHg	3.8
Odour threshold (ppm)	100	POCP	

Aqueous effluent

Solubility in water (25°C %w/w)	Total
Solubility of water in (25°C %w/w)	Total
Log_{10} activated carbon partition	
Log_{10} partition in octanol/water (w/w)	–0.74
Biological oxygen demand w/w (days)	0.9
Theoretical oxygen demand w/w	1.86

Vapour pressure equation constants (Log_{10}, mmHg)

Antoine equation	A	7.10850
	B	1537.78
	C	210.390
Cox chart	A	
	B	

Solvent properties

Solubility parameter	12.1	Kauri butanol value	
Dipole (D)	3.8	Evaporation time (ether = 1)	120
Dielectric constant (20°C)	36.7	Evaporation time (BuAc = 1)	0.17
Polarity (water 100)	40.4		

Thermal information

Latent heat (cal/mol)	10074
Nett heat of combustion (kcal/gmol)	423
Specific heat (cal/mol/°C)	36
Critical pressure (MN/m^2)	4.48
Critical temperature (K)	647
Latent heat of fusion (cal/mol)	
Van der Waals' volume	3.09
Van der Waals' surface area	2.74
Molar volume	77.43

Solute	Azeotrope X% w/w	Azeotrope °C	Reference	Solute γ^{∞}	Reference	Partition coefficient	Reference
Hydrocarbons							
n-Pentane				12.8	1x/1/58		
n-Hexane				17.2	6c/332		
n-Heptane	5	97		8.3	6x/98		
n-Octane				18.6	1x/1/61		
n-Nonane				22.0	1x/3/1005		
n-Decane				25.3	1x/3/1005		
2,2,4-TMP				30.1	1x/1/61		
Cyclohexane				5.4	1x/1/59		
Benzene	None		5893	1.2	1x/1/59		
Toluene	None		5893a	2.0	1x/1/60		
Ethylbenzene	15	134		2.2	1x/3/1003		
Xylenes	20	136	5894	2.7	7/481		
C_9 Aromatics							
Tetralin							
Alcohols							
Methanol	None			0.6	2a/115		
Ethanol	None			0.6	2c/371		
n-Propanol							
i-Propanol							
n-Butanol							
i-Butanol							
s-Butanol							
n-Amyl alc.							
i-Amyl alc.							
Cyclohexanol							
1-Octanol				1.2	2f/527		
Ethanediol				1.0	2b/8		
DEG							
1,2-Propanediol							
Glycol ethers							
PGME							
EGME							
EEE							
EGBE							
Chlorinated							
MDC				0.9	8/265	0.4	
Chloroform					0.29		
Carbon tet.				1.8	8/117		
1,2-EDC							
1,1,1-TCA							
TCE						0.04	
Perk.						0.02	
MCB							

Miscellaneous solvents

Solute	Azeotrope X% w/w °C	Azeotrope Reference	Solute γ^{∞}	Reference	Partition coefficient	Reference
Ketones						
Acetone			0.9	3+4/164		
MEK			1.3	3b/289		
MIBK						
Cyclohexanone			1.2	3b/500		
NMP						
Acetophenone						
Ethers						
Diethyl ether						
DIPE						
Dibutyl ether						
MTBE						
1,4-Dioxane			1.3	3+4/454		
THF						
Esters						
Me acetate			1.6	1x/3/999		
Et acetate						
i-Propyl acetate						
n-Butyl acetate						
Cellosolve acetate						
Miscellaneous						
DMF	–	–	–	–	–	–
DMAc						
DMSO			1.2	8/407		
Sulfolane						
CS_2			4.4	1x/1/57		
Acetic acid						
Aniline	Azeo					
Nitrobenzene						
Morpholine						
Pyridine						
2-Nitropropane						
Acetonitrile	None	2765	0.3	8/428		
Furfuraldehyde						
Phenol						
Water	None		1.08	1/276		

Dimethylacetamide

Alternative names
DMAc

Reference codes

CAS number	127 19 5	Hazchem code	
UN number		EPA code	

Physical properties

Molecular weight	87	Cubic expansion coeff (per °C × 10^3)	0.95
Empirical formula	$C_4H_9N_1O_1$	Surface tension (@20°C dyn/cm)	34
Boiling point (°C)	166	Absolute viscosity (@25°C cP)	0.92
Freezing point (°C)	–20	Refractive index (25°C)	1.436
Specific gravity (20/4)	0.945		

Fire hazards

Flash point (closed cup °C)	70	Lower explosive limit (ppm)	15000
Autoignition temperature (°C)	491	Upper explosive limit (ppm)	115000
Electrical conductivity			

Health hazards

IDLH (ppm)	400	Vapour concentration @21°C ppm	1316
OES-TWA	10	Vapour density (relative to air)	3.02
OES-STEL	20	Vapour pressure @21°C mmHg	1.0
Odour threshold (ppm)	50	POCP	

Aqueous effluent

Solubility in water (25°C %w/w)	Total
Solubility of water in (25°C %w/w)	Total
Log_{10} activated carbon partition	
Log_{10} partition in octanol/water (w/w)	–0.77
Biological oxygen demand w/w (days)	
Theoretical oxygen demand w/w	

Vapour pressure equation constants (Log_{10}, mmHg)

Antoine equation	A	7.76228
	B	1889.1
	C	221.00
Cox chart	A	
	B	

Solvent properties

Solubility parameter	11.0	Kauri butanol value	
Dipole (D)	3.8	Evaporation time (ether = 1)	172
Dielectric constant (20°C)	37.8	Evaporation time (BuAc = 1)	0.14
Polarity (water 100)	40.1		

Thermal information

Latent heat (cal/mol)	10360
Nett heat of combustion (kcal/gmol)	569
Specific heat (cal/mol/°C)	42
Critical pressure (MN/m^2)	4.08
Critical temperature (K)	658
Latent heat of fusion (cal/mol)	
Van der Waals' volume	3.53
Van der Waals' surface area	2.97
Molar volume	92.1

Solute	Azeotrope X% w/w	°C	Reference	Solute γ^∞	Reference	Partition coefficient	Reference
Hydrocarbons							
n-Pentane				8.3	1x/1/118		
n-Hexane				17.0	1x/1/118		
n-Heptane				20.0	1x/1/118		
n-Octane				13.0	1x/3/1168		
n-Nonane							
n-Decane							
2,2,4-TMP							
Cyclohexane							
Benzene				2.3	1x/1/118		
Toluene				3.0	1x/1/118		
Ethylbenzene				1.6	1x/3/1168		
Xylenes				1.6	1x/3/1168		
C_9 Aromatics							
Tetralin							
Alcohols							
Methanol				0.5	1x/3/1065		
Ethanol				0.5	1x/3/1068		
n-Propanol				0.3	2e/454		
i-Propanol							
n-Butanol							
i-Butanol							
s-Butanol							
n-Amyl alc.							
i-Amyl alc.							
Cyclohexanol							
1-Octanol							
Ethanediol							
DEG							
1,2-Propanediol							
Glycol ethers							
PGME							
EGME							
EEE							
EGBE							
Chlorinated							
MDC							
Chloroform							
Carbon tet.							
1,2-EDC							
1,1,1-TCA							
TCE							
Perk.							
MCB							

Miscellaneous solvents

Solute	Azeotrope X% w/w	°C	Reference	Solute γ^∞	Reference	Partition coefficient	Reference
Ketones							
Acetone				1.1	1x/3/1168		
MEK				1.1	1x/3/1168		
MIBK							
Cyclohexanone							
NMP							
Acetophenone							
Ethers							
Diethyl ether							
DIPE							
Dibutyl ether							
MTBE							
1,4-Dioxane				1.2	1x/3/1168		
THF							
Esters							
Me acetate				1.3	1x/3/1168		
Et acetate				1.3	1x/3/1168		
i-Propyl acetate							
n-Butyl acetate							
Cellosolve acetate							
Miscellaneous							
DMF							
DMAc	–		–	–	–	–	–
DMSO							
Sulfolane							
CS_2							
Acetic acid	79	171	3133	0.2	5/115		
Aniline							
Nitrobenzene							
Morpholine							
Pyridine							
2-Nitropropane							
Acetonitrile							
Furfuraldehyde							
Phenol							
Water	None			1.0	1a/402		

Dimethylsulphoxide

Alternative names
DMSO, sulfinyl-bis-methane, DIMSO

Reference codes			
CAS number	67 68 5	Hazchem code	
UN number		EPA code	

Physical properties			
Molecular weight	78	Cubic expansion coeff (per °C × 10^3)	1.0
Empirical formula	$C_2H_6O_1S_1$	Surface tension (@20°C dyn/cm)	43.7
Boiling point (°C)	189	Absolute viscosity (@25°C cP)	2.0
Freezing point (°C)	+18.5	Refractive index (25°C)	1.476
Specific gravity (20/4)	1.101		

Fire hazards			
Flash point (closed cup °C)	95	Lower explosive limit (ppm)	30000
Autoignition temperature (°C)	255	Upper explosive limit (ppm)	420000
Electrical conductivity	2E–9		

Health hazards			
IDLH (ppm)	1000	Vapour concentration @21°C ppm	650
OES-TWA		Vapour density (relative to air)	2.7
OES-STEL		Vapour pressure @21°C mmHg	0.7
Odour threshold (ppm)		POCP	

Aqueous effluent	
Solubility in water (25°C %w/w)	Total
Solubility of water in (25°C %w/w)	Total
Log_{10} activated carbon partition	
Log_{10} partition in octanol/water (w/w)	–2.03
Biological oxygen demand w/w (days)	
Theoretical oxygen demand w/w	2.05

Vapour pressure equation constants (Log_{10}, mmHg)		
Antoine equation	A	6.88076
	B	1541.52
	C	191.797
Cox chart	A	
	B	

Solvent properties			
Solubility parameter	13.0	Kauri butanol value	
Dipole (D)	3.96	Evaporation time (ether = 1)	1500
Dielectric constant (20°C)	46.6	Evaporation time (BuAc = 1)	
Polarity (water 100)	44.4		

Thermal information	
Latent heat (cal/mol)	12636
Nett heat of combustion (kcal/gmol)	441
Specific heat (cal/mol/°C)	36
Critical pressure (MN/m^2)	
Critical temperature (K)	
Latent heat of fusion (cal/mol)	3221
Van der Waals' volume	2.83
Van der Waals' surface area	2.47
Molar volume	71.3

Solute	Azeotrope X% w/w	°C	Azeotrope Reference	Solute γ^{∞}	Reference	Partition coefficient	Reference
Hydrocarbons							
n-Pentane				25.9	1x/1/29		
n-Hexane				38.6	1x/1/30		
n-Heptane				33.4	1x/1/31		
n-Octane				43.5	1x/1/31		
n-Nonane				55.9	1x/1/31		
n-Decane				70.7	1x/1/32		
2,2,4-TMP				61.2	1x/1/31		
Cyclohexane				15.5	1x/1/30		
Benzene	None		4184	2.7	7/169		
Toluene				4.1	7/386		
Ethylbenzene				4.6	1x/1/31		
Xylenes				7.5	1x/3/969		
C_9 Aromatics							
Tetralin							
Alcohols							
Methanol				0.4	2c/62		
Ethanol				0.8	1x/3/967		
n-Propanol							
i-Propanol				1.5	2f/39		
n-Butanol	None		4183a	0.7	2f/131		
i-Butanol	None			0.9	2b/275		
s-Butanol							
n-Amyl alc.							
i-Amyl alc.							
Cyclohexanol							
1-Octanol							
Ethanediol	None			0.2	2b/7		
DEG							
1,2-Propanediol							
Glycol ethers							
PGME							
EGME							
EEE							
EGBE							
Chlorinated							
MDC	None			0.6	8/264		
Chloroform	None			0.7	8/229		
Carbon tet.	None			4.0	8/107		
1,2-EDC							
1,1,1-TCA							
TCE							
Perk.							
MCB							

Miscellaneous solvents

Solute	Azeotrope		Solute γ^∞	Reference	Partition coefficient	Reference
	X% w/w °C	Reference				
Ketones						
Acetone	None		1.8	3b/80		
MEK			2.5	1x/3/967		
MIBK						
Cyclohexanone						
NMP						
Acetophenone						
Ethers						
Diethyl ether						
DIPE						
Dibutyl ether						
MTBE						
1,4-Dioxane	None		1.5	3+4/450		
THF	None		2.5	3+4/433		
Esters						
Me acetate			3.2	1x/3/967		
Et acetate			3.3	5/461		
i-Propyl acetate						
n-Butyl acetate						
Cellosolve acetate						
Miscellaneous						
DMF	None		1.1	8/407		
DMAc						
DMSO	–	–	–	–	–	–
Sulfolane						
CS_2						
Acetic acid						
Aniline						
Nitrobenzene						
Morpholine						
Pyridine						
2-Nitropropane						
Acetonitrile						
Furfuraldehyde						
Phenol						
Water	None	243	0.5	1/119		

Sulfolane

Alternative names
(Cyclo)tetramethylene sulphone, thiolane-1,1-dioxide

Reference codes			
CAS number	126 33 0	Hazchem code	
UN number		EPA code	

Physical properties			
Molecular weight	120	Cubic expansion coeff (per °C × 10^3)	0.7
Empirical formula	$C_4H_8O_2S$	Surface tension (@30°C dyn/cm)	35.5
Boiling point (°C)	285	Absolute viscosity (@30°C cP)	10.3
Freezing point (°C)	+27.4	Refractive index (30°C)	1.471
Specific gravity (20/4)	1.26		

Fire hazards			
Flash point (closed cup °C)	177	Lower explosive limit (ppm)	54000
Autoignition temperature (°C)		Upper explosive limit (ppm)	
Electrical conductivity			

Health hazards			
IDLH (ppm)		Vapour concentration @21°C ppm	65.8
OES-TWA		Vapour density (relative to air)	4.17
OES-STEL		Vapour pressure @21°C mmHg	0.05
Odour threshold (ppm)		POCP	

Aqueous effluent			
Solubility in water (25°C %w/w)		Total	
Solubility of water in (25°C %w/w)		Total	
Log_{10} activated carbon partition			
Log_{10} partition in octanol/water (w/w)			
Biological oxygen demand w/w (days)			
Theoretical oxygen demand w/w			

Vapour pressure equation constants (Log_{10}, mmHg)		
Antoine equation	A	7.40800
	B	2255.469
	C	211.393
Cox chart	A	
	B	

Solvent properties			
Solubility parameter		Kauri butanol value	
Dipole (D)	4.69	Evaporation time (ether = 1)	
Dielectric constant (20°C)	44	Evaporation time (BuAc = 1)	
Polarity (water 100)	41		

Thermal information	
Latent heat (cal/mol)	14720
Nett heat of combustion (kcal/gmol)	595
Specific heat (cal/mol/°C)	55
Critical pressure (MN/m^2)	5.32
Critical temperature (K)	801
Latent heat of fusion (cal/mol)	1063
Van der Waals' volume	4.04
Van der Waals' surface area	3.20
Molar volume	95.3

Solute	Azeotrope X% w/w	Azeotrope °C	Reference	Solute γ^∞	Reference	Partition coefficient	Reference
Hydrocarbons							
n-Pentane				33.2	1x/1/109		
n-Hexane				48.2	1x/1/111		
n-Heptane	Azeo			51.4	1x/1/112		
n-Octane				66.0	1x/1/113		
n-Nonane				87.3	1x/1/114		
n-Decane				115.0	1x/1/114		
2,2,4-TMP				53.2	1x/1/113		
Cyclohexane				19.3	1x/1/110		
Benzene				2.7	7/191		
Toluene	Azeo			1.5	7/399		
Ethylbenzene				4.9	1x/1/112		
Xylenes				5.1	1x/1/112		
C_9 Aromatics				6.7	1x/3/1063		
Tetralin							
Alcohols							
Methanol				0.8	2c/125		
Ethanol				3.3	2c/344		
n-Propanol				3.5	1x/3/1054		
i-Propanol				4.9	2d/53		
n-Butanol				4.6	1x/3/1055		
i-Butanol				4.3	1x/3/1055		
s-Butanol							
n-Amyl alc.							
i-Amyl alc.							
Cyclohexanol							
1-Octanol							
Ethanediol							
DEG							
1,2-Propanediol							
Glycol ethers							
PGME							
EGME				1.7	2f/103		
EEE							
EGBE							
Chlorinated							
MDC	None			0.9	8/266		
Chloroform				1.1	1x/3/1053		
Carbon tet.				5.0	1x/3/1053		
1,2-EDC				1.3	1x/3/1053		
1,1,1-TCA							
TCE							
Perk.							
MCB							

Miscellaneous solvents

Solute	Azeotrope						
	X% °C w/w	Reference	Solute γ^{∞}	Reference	Partition coefficient	Reference
Ketones						
Acetone			1.5	1x/3/1054		
MEK			2.1	1x/3/1055		
MIBK			3.7	1x/3/1058		
Cyclohexanone						
NMP						
Acetophenone						
Ethers						
Diethyl ether						
DIPE						
Dibutyl ether						
MTBE						
1,4-Dioxane			3.3	1x/3/1055		
THF						
Esters						
Me acetate			1.7	1x/3/1054		
Et acetate			2.8	1x/3/1055		
i-Propyl acetate						
n-Butyl acetate						
Cellosolve acetate						
Miscellaneous						
DMF						
DMAc						
DMSO						
Sulfolane	–	–	–	–	–	–
CS_2						
Acetic acid						
Aniline						
Nitrobenzene						
Morpholine						
Pyridine						
2-Nitropropane						
Acetonitrile						
Furfuraldehyde						
Phenol						
Water	None		2.1	1x/3/1065		

Carbon disulphide

Alternative names
Carbon bisulphide

Reference codes

CAS number	75 15 0	Hazchem code	
UN number	1131	EPA code	P022

Physical properties

Molecular weight	76	Cubic expansion coeff (per °C × 10^3)	1.4
Empirical formula	C_1S_2	Surface tension (@20°C dyn/cm)	32
Boiling point (°C)	46	Absolute viscosity (@25°C cP)	0.36
Freezing point (°C)	–111	Refractive index (25°C)	1.628
Specific gravity (20/4)	1.26		

Fire hazards

Flash point (closed cup °C)	–30	Lower explosive limit (ppm)	13000
Autoignition temperature (°C)	102	Upper explosive limit (ppm)	500000
Electrical conductivity	1.0E–16		

Health hazards

Idlh (ppm)	500	vapour concentration @21°c ppm	685000
OES-TWA	10	Vapour density (relative to air)	2.7
OES-STEL		Vapour pressure @21°C mmHg	309
Odour threshold (ppm)	0.2	POCP	

Aqueous effluent

Solubility in water (25°C %w/w)	0.21
Solubility of water in (25°C %w/w)	0.014
Log_{10} activated carbon partition	
Log_{10} partition in octanol/water (w/w)	2.0
Biological oxygen demand w/w (days)	
Theoretical oxygen demand w/w	

Vapour pressure equation constants (Log_{10}, mmHg)

Antoine equation	A	6.94279
	B	1169.11
	C	241.59
Cox chart	A	
	B	

Solvent properties

Solubility parameter	10.0	Kauri butanol value	
Dipole (D)	0	Evaporation time (ether = 1)	1.8
Dielectric constant (20°C)	2.64	Evaporation time (BuAc = 1)	10.9
Polarity (water 100)	6.5		

Thermal information

Latent heat (cal/mol)	6460
Nett heat of combustion (kcal/gmol)	246
Specific heat (cal/mol/°C)	18
Critical pressure (MN/m^2)	7.62
Critical temperature (K)	546
Latent heat of fusion (cal/mol)	1050
Van der Waals' volume	2.06
Van der Waals' surface area	1.65
Molar volume	60.65

Solute	Azeotrope X% w/w	°C	Reference	Solute γ^∞	Reference	Partition coefficient	Reference
Hydrocarbons							
n-Pentane	11	36	1256				
n-Hexane	None		1274				
n-Heptane	None		1278				
n-Octane				2.0	1x/3/938		
n-Nonane							
n-Decane				1.5	6c/571		
2,2,4-TMP							
Cyclohexane	None		1269	0.4	6a/154		
Benzene	None		1265	1.4	7/100		
Toluene	None		1276	1.1	7/361		
Ethylbenzene							
Xylenes							
C_9 Aromatics							
Tetralin							
Alcohols							
Methanol	71	40	1175	6.3	2a/35		
Ethanol	91	43	1189	84.4	2a/281	16.4	P385
n-Propanol	95	46	1209	13.7	2c/417		
i-Propanol	92	44	1208				
n-Butanol	None		1233	11.3	2f/120		
i-Butanol	None		1236				
s-Butanol	None		1260	24.8	2d/239		
n-Amyl alc.	None		1257	8.7	2f/371		
i-Amyl alc.							
Cyclohexanol							
1-Octanol							
Ethanediol							
DEG							
1,2-Propanediol							
Glycol ethers							
PGME							
EGME							
EEE							
EGBE							
Chlorinated							
MDC	35	36	1170				
Chloroform	None		1169	1.4	8/214		
Carbon tet.	None		1085	1.4	8/76		
1,2-EDC							
1,1,1-TCA							
TCE							
Perk.				1.7	8/318		
MCB							

Miscellaneous solvents

Solute	Azeotrope X% w/w	°C	Reference	Solute γ^∞	Reference	Partition coefficient	Reference
Ketones							
Acetone	67	39	1194	7.0	3+4/132	0.78	P506
MEK	84	46	1216	4.4	1x/3/938		
MIBK	None		1272				
Cyclohexanone							
NMP							
Acetophenone							
Ethers							
Diethyl ether	13	34	1235	2.0	3+4/495		
DIPE							
Dibutyl ether							
MTBE							
1,4-Dioxane				3.3	3+4/446		
THF							
Esters							
Me acetate	70	40	1198	6.6	5/349	0.15	P520
Et acetate	93	46	1220				
i-Propyl acetate	None		1251				
n-Butyl acetate							
Cellosolve acetate							
Miscellaneous							
DMF							
DMAc							
DMSO							
Sulfolane							
CS_2	–		–	–	–	–	–
Acetic acid	None		1180			127.2	P333
Aniline							
Nitrobenzene							
Morpholine							
Pyridine							
2-Nitropropane							
Acetonitrile	88			26.6	8/320		
Furfuraldehyde							
Phenol						0.43	P1663
Water	97	43	207				

Acetic acid

Alternative names
Ethanoic acid, glacial acetic acid

Reference codes

CAS number	64 19 7	Hazchem code	
UN number	1842	EPA code	

Physical properties

Molecular weight	60	Cubic expansion coeff (per °C × 10^3)	1.14
Empirical formula	$C_2H_4O_2$	Surface tension (@20°C dyn/cm)	27.4
Boiling point (°C)	118	Absolute viscosity (@25°C cP)	1.13
Freezing point (°C)	+17	Refractive index (25°C)	1.370
Specific gravity (20/4)	1.051		

Fire hazards

Flash point (closed cup °C)	40	Lower explosive limit (ppm)	54000
Autoignition temperature (°C)	427	Upper explosive limit (ppm)	160000
Electrical conductivity	6E–9		

Health hazards

IDLH (ppm)	1000	Vapour concentration @21°C ppm	17400
OES-TWA	10	Vapour density (relative to air)	2.08
OES-STEL	15	Vapour pressure @21°C mmHg	13
Odour threshold (ppm)	2	POCP	

Aqueous effluent

Solubility in water (25°C %w/w)	Total
Solubility of water in (25°C %w/w)	Total
Log_{10} activated carbon partition	1.80
Log_{10} partition in octanol/water (w/w)	–0.24
Biological oxygen demand w/w (days)	
Theoretical oxygen demand w/w	1.07

Vapour pressure equation constants (Log_{10}, mmHg)

Antoine equation	A	7.5596
	B	1644.05
	C	233.5
Cox chart	A	7.4565
	B	1592.4

Solvent properties

Solubility parameter	10.1	Kauri butanol value	
Dipole (D)	1.7	Evaporation time (ether = 1)	
Dielectric constant (20°C)	6.2	Evaporation time (BuAc = 1)	
Polarity (water 100)	64.8		

Thermal information

Latent heat (cal/mol)	5800
Nett heat of combustion (kcal/gmol)	188
Specific heat (cal/mol/°C)	29.4
Critical pressure (MN/m^2)	5.8
Critical temperature (K)	595
Latent heat of fusion (cal/mol)	2800
Van der Waals' volume	2.20
Van der Waals' surface area	2.07
Molar volume	57.5

Solute	Azeotrope X% w/w	Azeotrope °C	Azeotrope Reference	Solute γ^∞	Reference	Partition coefficient	Reference
Hydrocarbons							
n-Pentane	None		3156				
n-Hexane	6	68	3184	9.0	5/152		
n-Heptane	25	93	3204	20	5/175		
n-Octane	53	105	3219	16	5/189		
n-Nonane	69	113	3230				
n-Decane	79	117	3237	31	5/191		
2,2,4-TMP							
Cyclohexane	10	79	3173	4.9	5/146		
Benzene	2	80	3163	3.3	5/127		
Toluene	34	105	3194	7.6	5/159		
Ethylbenzene	66	115	3206	7.9	5/178		
Xylenes	73	116	3208	5.6	5/181		
C_9 Aromatics							
Tetralin							
Alcohols							
Methanol	None		1933	0.9	2a/48		
Ethanol	None		3090	0.7	2c/293		
n-Propanol	None		3109	0.9	2a/525		
i-Propanol				0.8	2d/84		
n-Butanol	43	120	3135	1.0	2d/158		
i-Butanol				1.1	2f/302		
s-Butanol				1.2	2f/221		
n-Amyl alc.							
i-Amyl alc.	16	133	3156a				
Cyclohexanol							
1-Octanol							
Ethanediol							
DEG							
1,2-Propanediol							
Glycol ethers							
PGME							
EGME							
EEE							
EGBE							
Chlorinated							
MDC				1.9	5/64		
Chloroform	None		1437	2.6	5/62		
Carbon tet.	2	76	1099	5.4	5/59		
1,2-EDC	None		2961	2.6	5/74		
1,1,1-TCA							
TCE	4	118	2282	3.3	5/72		
Perk.	39	107	2158				
MCB	59	115	3160				

Miscellaneous solvents

Solute	Azeotrope X% w/w	°C	Reference	Solute γ^∞	Reference	Partition coefficient	Reference
Ketones							
Acetone	None		3101	1.0	3+4/148		
MEK	None		3117	1.1	3+4/269		
MIBK	None		3175	1.5	3+4/345		
Cyclohexanone							
NMP							
Acetophenone							
Ethers							
Diethyl ether	None		3136	1.9	3+4/502		
DIPE	None		3185	2.4	3+4/544		
Dibutyl ether	None		3220				
MTBE							
1,4-Dioxane	78	119	3119	0.9	3+4/448		
THF							
Esters							
Me acetate	None		3101a	1.1	5/82		
Et acetate	None		3120	1.2	5/104		
i-Propyl acetate	None		3147	1.6	5/123		
n-Butyl acetate	None		3177	1.8	5/147		
Cellosolve acetate							
Miscellaneous							
DMF							
DMAc	21	171	3133	0.1	5.115		
DMSO							
Sulfolane							
CS_2	None		1180				
Acetic acid	–		–	–	–	–	–
Aniline	None		3164				
Nitrobenzene	None		3162				
Morpholine							
Pyridine	51	138	3139	0.04	5/118		
2-Nitropropane							
Acetonitrile	None		2758a				
Furfuraldehyde	None		3138				
Phenol							
Water	None		231	1.4	1/102		

Aniline

Alternative names
Aminobenzene, benzeneamine, phenylamine

Reference codes

CAS number	62 53 3	Hazchem code	
UN number	1547	EPA code	U012

Physical properties

Molecular weight	93	Cubic expansion coeff (per °C × 10^3)	0.9
Empirical formula	$C_6H_7N_1$	Surface tension (@20°C dyn/cm)	45.5
Boiling point (°C)	184	Absolute viscosity (@25°C cP)	4.4
Freezing point (°C)	–6	Refractive index (25°C)	1.583
Specific gravity (20/4)	1.022		

Fire hazards

Flash point (closed cup °C)	76	Lower explosive limit (ppm)	13000
Autoignition temperature (°C)	770	Upper explosive limit (ppm)	110000
Electrical conductivity	2.4E–8		

Health hazards

IDLH (ppm)	100	Vapour concentration @21°C ppm	380
OES-TWA	0.5	Vapour density (relative to air)	3.23
OES-STEL		Vapour pressure @21°C mmHg	0.26
Odour threshold (ppm)	0.5	POCP	

Aqueous effluent

Solubility in water (25°C %w/w)	3.5
Solubility of water in (25°C %w/w)	5.1
Log_{10} activated carbon partition	2.78
Log_{10} partition in octanol/water (w/w)	+0.94
Biological oxygen demand w/w (days)	1.50 (5)
Theoretical oxygen demand w/w	2.67

Vapour pressure equation constants (Log_{10}, mmHg)

Antoine equation	A	7.46441
	B	1840.79
	C	216.92
Cox chart	A	7.55756
	B	1936.2

Solvent properties

Solubility parameter	10.3	Kauri butanol value	
Dipole (D)	1.56	Evaporation time (ether = 1)	
Dielectric constant (30°C)	6.7	Evaporation time (BuAc = 1)	
Polarity (water 100)	42.0		

Thermal information

Latent heat (cal/mol)	10602
Nett heat of combustion (kcal/gmol)	773
Specific heat (cal/mol/°C)	48
Critical pressure (MN/m^2)	5.30
Critical temperature (K)	699
Latent heat of fusion (cal/mol)	2519
Van der Waals' volume	3.72
Van der Waals' surface area	2.82
Molar volume	91.53

Solute	Azeotrope X% w/w	Azeotrope °C	Azeotrope Reference	Solute γ^{∞}	Reference	Partition coefficient	Reference
Hydrocarbons							
n-Pentane				17.4	1x/1/240		
n-Hexane	None		11151	6.8	6a/580		
n-Heptane	None		11179	11.0	6b/161		
n-Octane	None		11197	38	1x/3/1194		
n-Nonane	13	149	11210	26	1x/1/242		
n-Decane	36	167	11230				
2,2,4-TMP	None		11196	6.0	6b/318		
Cyclohexane	None		11148	4.1	6a/255		
Benzene	None		10850	1.4	7/263		
Toluene	None		11167	1.6	7/426		
Ethylbenzene	None		11183	2.3	7/474		
Xylenes	None		11185	3.1	1x/3/1194		
C_9 Aromatics							
Tetralin							
Alcohols							
Methanol				2.2	1x/3/1192		
Ethanol	None		4075	3.3	2a/427		
n-Propanol							
i-Propanol							
n-Butanol	None		8138	2.3	1x/3/1193		
i-Butanol							
s-Butanol				2.9	2b/265		
n-Amyl alc.							
i-Amyl alc.	None		9824				
Cyclohexanol	None		11149	1.6	2b/395		
1-Octanol	83	184	11200				
Ethanediol	76	181	4242	3.6	2b/16		
DEG							
1,2-Propanediol	57	186	6655				
Glycol ethers							
PGME	None		9963				
EGME							
EEE							
EGBE	None		11153				
Chlorinated							
MDC				1.4	1x/1/239		
Chloroform				1.5	1x/1/239		
Carbon tet.	None			3.8	8/174		
1,2-EDC							
1,1,1-TCA							
TCE							
Perk.							
MCB	None		10511	1.8	8/527		

Miscellaneous solvents

Solute	Azeotrope X% w/w	°C	Reference	Solute γ^{∞}	Reference	Partition coefficient	Reference
Ketones							
Acetone	None			1.3	3b/183		
MEK				0.6	1x/3/1192		
MIBK							
Cyclohexanone	None		11144				
NMP							
Acetophenone	None		11182				
Ethers							
Diethyl ether							
DIPE							
Dibutyl ether	None		11198				
MTBE							
1,4-Dioxane				0.4	1x/3/1192		
THF							
Esters							
Me acetate							
Et acetate				1.4	1x/1/239		
i-Propyl acetate							
n-Butyl acetate							
Cellosolve acetate							
Miscellaneous							
DMF	Azeo?			0.3	8/428		
DMAc							
DMSO							
Sulfolane							
CS_2				3.5	1x/1/239		
Acetic acid	None		3164				
Aniline	–		–	–	–	–	–
Nitrobenzene	3	180	10704	6.4	8/540		
Morpholine							
Pyridine	Azeo?			?	8/500	0.05	V3/233
2-Nitropropane							
Acetonitrile	None			0.9	8/385		
Furfuraldehyde							
Phenol						0.01	V3/290
Water	25	81	488	4.5	1/499		

Nitrobenzene

Alternative names
Oil of mirbane, nitrobenzol

Reference codes

CAS number	98 95 3	Hazchem code	2Y
UN number	1662	EPA code	U169

Physical properties

Molecular weight	123	Cubic expansion coeff (per °C × 10^3)	0.96
Empirical formula	$C_6H_6N_1O_2$	Surface tension (@20°C dyn/cm)	43.9
Boiling point (°C)	211	Absolute viscosity (@25°C cP)	1.80
Freezing point (°C)	+6	Refractive index (25°C)	1.550
Specific gravity (20/4)	1.204		

Fire hazards

Flash point (closed cup °C)	88	Lower explosive limit (ppm)	18000
Autoignition temperature (°C)	496	Upper explosive limit (ppm)	
Electrical conductivity	2E–10		

Health hazards

IDLH (ppm)	200	Vapour concentration @21°C ppm	272
OES-TWA	1	Vapour density (relative to air)	4.27
OES-STEL	2	Vapour pressure @21°C mmHg	0.21
Odour threshold (ppm)	6	POCP	

Aqueous effluent

Solubility in water (25°C %w/w)	0.19
Solubility of water in (25°C %w/w)	
Log_{10} activated carbon partition	
Log_{10} partition in octanol/water (w/w)	+1.86
Biological oxygen demand w/w (days)	
Theoretical oxygen demand w/w	1.82

Vapour pressure equation constants (Log_{10}, mmHg)

Antoine equation	A	7.13043
	B	1751.36
	C	201.34
Cox chart	A	
	B	

Solvent properties

Solubility parameter	10.0	Kauri butanol value	
Dipole (D)	4.0	Evaporation time (ether = 1)	
Dielectric constant (20°C)	34.8	Evaporation time (BuAc = 1)	
Polarity (water 100)	32.4		

Thermal information

Latent heat (cal/mol)	10455
Nett heat of combustion (kcal/gmol)	706
Specific heat (cal/mol/°C)	44
Critical pressure (MN/m^2)	4.82
Critical temperature (K)	720
Latent heat of fusion (cal/mol)	2768
Van der Waals' volume	4.08
Van der Waals' surface area	3.10
Molar volume	102.7

Solute	Azeotrope X% w/w	°C	Reference	Solute γ^{∞}	Reference	Partition coefficient	Reference
Hydrocarbons							
n-Pentane	None		9740	7.0	1x/1/222		
n-Hexane	None		10708	8.7	6a/532		
n-Heptane				6.7	1x/1/223		
n-Octane				3.9	6b/241		
n-Nonane				8.0	1x/1/223		
n-Decane							
2,2,4-TMP				11.8	1x/1/223		
Cyclohexane				9.6	6a/203		
Benzene	None		10703	1.1	7/253		
Toluene	None		10718	1.5	7/422		
Ethylbenzene				1.6	7/470		
Xylenes				1.8	1x/1/223		
C_9 Aromatics							
Tetralin							
Alcohols							
Methanol	None		2065	10.4	1x/1/220	5.9	P165
Ethanol	None		4072	10.7	1x/1/220	3.5	V2/349
n-Propanol							
i-Propanol							
n-Butanol	None		8135				
i-Butanol							
s-Butanol	None		8231	16.0	2f/226		
n-Amyl alc.						0.02	V3/257
i-Amyl alc.							
Cyclohexanol							
1-Octanol	None		10754				
Ethanediol	41	186	4238				
DEG	90	210	8518				
1,2-Propanediol							
Glycol ethers							
PGME							
EGME							
EEE							
EGBE	None		10710				
Chlorinated							
MDC				1.0	1x/1/220		
Chloroform	None		1485	1.0	1/1/220		
Carbon tet.	None		1153	1.8	8/168		
1,2-EDC				1.1	1x/1/220		
1,1,1-TCA							
TCE							
Perk.							
MCB	None		10508	1.1	1x/3/1180		

Miscellaneous solvents

Solute	Azeotrope X% w/w	Azeotrope °C	Reference	Solute γ^{∞}	Reference	Partition coefficient	Reference
Ketones							
Acetone				1.2	1x/1/221		
MEK				1.1	3b/316		
MIBK							
Cyclohexanone							
NMP							
Acetophenone	None		10733				
Ethers							
Diethyl ether	None		8297				
DIPE							
Dibutyl ether							
MTBE							
1,4-Dioxane				0.8	1x/3/1180		
THF							
Esters							
Me acetate							
Et acetate				1.4	1x/1/221		
i-Propyl acetate							
n-Butyl acetate							
Cellosolve acetate							
Miscellaneous							
DMF							
DMAc							
DMSO							
Sulfolane							
CS_2				2.6	1x/1/220		
Acetic acid	None		3162			3.9	P311
Aniline	97	180	10704	1.0	8/540	0.01	V3/260
Nitrobenzene	–		–	–	–	–	–
Morpholine							
Pyridine							
2-Nitropropane							
Acetonitrile				1.7	1x/1/220		
Furfuraldehyde							
Phenol						0.02	V3/259
Water	12	99	485				

Morpholine

Alternative names
Tetrahydro. p. oxazine, diethyeneimide oxide, tetrahydro-1,4-isoxazine

Reference codes

CAS number	110 91 8	Hazchem code	2P
UN number	2054	EPA code	

Physical properties

Molecular weight	87	Cubic expansion coeff (per °C × 10^3)	1.1
Empirical formula	$C_4H_9N_1O_1$	Surface tension (@20°C dyn/cm)	37.5
Boiling point (°C)	129	Absolute viscosity (@25°C cP)	2.2
Freezing point (°C)	–5	Refractive index (25°C)	1.455
Specific gravity (20/4)	1.00		

Fire hazards

Flash point (closed cup °C)	38	Lower explosive limit (ppm)	18000
Autoignition temperature (°C)	310	Upper explosive limit (ppm)	108000
Electrical conductivity	6E–10		

Health hazards

IDLH (ppm)		Vapour concentration @21°C ppm	10500
OES-TWA	20	Vapour density (relative to air)	3.0
OES-STEL	30	Vapour pressure @21°C mmHg	7.9
Odour threshold (ppm)	1	POCP	

Aqueous effluent

Solubility in water (25°C %w/w)	Total
Solubility of water in (25°C %w/w)	Total
Log_{10} activated carbon partition	
Log_{10} partition in octanol/water (w/w)	–1.08
Biological oxygen demand w/w (days)	
Theoretical oxygen demand w/w	2.11

Vapour pressure equation constants (Log_{10}, mmHg)

Antoine equation	A	7.1603
	B	1447.7
	C	210.0
Cox chart	A	
	B	

Solvent properties

Solubility parameter		Kauri butanol value	
Dipole (D)	1.56	Evaporation time (ether = 1)	
Dielectric constant (20°C)	7.4	Evaporation time (BuAc = 1)	
Polarity (water 100)	31.8		

Thermal information

Latent heat (cal/mol)	9510
Nett heat of combustion (kcal/gmol)	
Specific heat (cal/mol/°C)	42
Critical pressure (MN/m^2)	5.5
Critical temperature (K)	618
Latent heat of fusion (cal/mol)	
Van der Waals' volume	3.47
Van der Waals' surface area	2.80
Molar volume	87.5

Solute	Azeotrope			Solute γ^{∞}	Reference	Partition coefficient	Reference
	X% w/w	°C	Reference				
Hydrocarbons							
n-Pentane							
n-Hexane				27.0	1x/1/123		
n-Heptane							
n-Octane							
n-Nonane							
n-Decane							
2,2,4-TMP							
Cyclohexane							
Benzene				3.3	1x/1/123		
Toluene							
Ethylbenzene				1.9	3+4/482		
Xylenes	None		8057				
C_9 Aromatics							
Tetralin							
Alcohols							
Methanol							
Ethanol				0.9	2c/345		
n-Propanol							
i-Propanol							
n-Butanol							
i-Butanol							
s-Butanol							
n-Amyl alc.							
i-Amyl alc.							
Cyclohexanol							
1-Octanol							
Ethanediol							
DEG							
1,2-Propanediol							
Glycol ethers							
PGME							
EGME							
EEE							
EGBE							
Chlorinated							
MDC							
Chloroform							
Carbon tet.							
1,2-EDC							
1,1,1-TCA							
TCE							
Perk.							
MCB							

Miscellaneous solvents

Solute	Azeotrope			Solute γ^∞	Reference	Partition coefficient	Reference
	X% w/w	°C	Reference				
Ketones							
Acetone							
MEK							
MIBK							
Cyclohexanone							
NMP							
Acetophenone							
Ethers							
Diethyl ether							
DIPE							
Dibutyl ether	73	127	8058				
MTBE							
1,4-Dioxane							
THF							
Esters							
Me acetate							
Et acetate							
i-Propyl acetate							
n-Butyl acetate							
Cellosolve acetate							
Miscellaneous							
DMF							
DMAc							
DMSO							
Sulfolane							
CS_2							
Acetic acid							
Aniline							
Nitrobenzene							
Morpholine	–		–	–	–	–	–
Pyridine							
2-Nitropropane							
Acetonitrile							
Furfuraldehyde							
Phenol							
Water	None		367	2.0	1a/327		

Pyridine

Alternative names

Reference codes

CAS number	110 86 1	Hazchem code	2WE
UN number	1252	EPA code	U196

Physical properties

Molecular weight	79	Cubic expansion coeff (per °C × 10^3)	1.0
Empirical formula	$C_5H_6N_1$	Surface tension (@20°C dyn/cm)	36.6
Boiling point (°C)	115	Absolute viscosity (@25°C cP)	0.88
Freezing point (°C)	–42	Refractive index (25°C)	1.507
Specific gravity (20/4)	0.983		

Fire hazards

Flash point (closed cup °C)	20	Lower explosive limit (ppm)	18000
Autoignition temperature (°C)	522	Upper explosive limit (ppm)	124000
Electrical conductivity	4E–8		

Health hazards

IDLH (ppm)	3600	Vapour concentration @21°C ppm	22000
OES-TWA	5	Vapour density (relative to air)	2.74
OES-STEL	10	Vapour pressure @21°C mmHg	16.6
Odour threshold (ppm)	0.03	POCP	

Aqueous effluent

Solubility in water (25°C %w/w)	Total
Solubility of water in (25°C %w/w)	Total
Log_{10} activated carbon partition	2.26
Log_{10} partition in octanol/water (w/w)	+0.64
Biological oxygen demand w/w (days)	1.47 (5)
Theoretical oxygen demand w/w	3.03

Vapour pressure equation constants (Log_{10}, mmHg)

Antoine equation	A	7.01328
	B	1356.93
	C	212.655
Cox chart	A	
	B	

Solvent properties

Solubility parameter	10.7	Kauri butanol value	
Dipole (D)	2.3	Evaporation time (ether = 1)	12.7
Dielectric constant (20°C)	12.9	Evaporation time (BuAc = 1)	
Polarity (water 100)	30.2		

Thermal information

Latent heat (cal/mol)	8374
Nett heat of combustion (kcal/gmol)	617
Specific heat (cal/mol/°C)	34
Critical pressure (MN/m^2)	5.64
Critical temperature (K)	620
Latent heat of fusion (cal/mol)	
Van der Waals' volume	3.00
Van der Waals' surface area	2.11
Molar volume	80.86

Solute	Azeotrope X% w/w	°C	Reference	Solute γ^∞	Reference	Partition coefficient	Reference
Hydrocarbons							
n-Pentane				7.2	1x/3/1092		
n-Hexane	None			4.8	1x/3/1092		
n-Heptane	25	96	8860	4.6	6b/116		
n-Octane	56	110	8867	3.9	6b/239		
n-Nonane	90	115	8870	4.2	6b/354		
n-Decane	None		8872	3.8	6b/386		
2,2,4-TMP	23	96	8868	10.3	6b/297		
Cyclohexane	None		8846	2.5	6a/177		
Benzene	None		8841	1.3	7/220		
Toluene	21	110	8858	1.5	7/406		
Ethylbenzene	None		8861				
Xylenes	None		8863	1.3	7/482		
C_9 Aromatics							
Tetralin							
Alcohols							
Methanol	None		2024	1.1	2a/183		
Ethanol	44	73	2760	1.1	2c/355		
n-Propanol	None		6469	0.9	2c/512		
i-Propanol				1.1	2d/57		
n-Butanol	30	119	8109	1.0	2b/166		
i-Butanol				0.4	2f/307		
s-Butanol	None		8217	0.9	2b/255		
n-Amyl alc.	None		8836				
i-Amyl alc.	None		8838				
Cyclohexanol							
1-Octanol							
Ethanediol	None		4215				
DEG							
1,2-Propanediol							
Glycol ethers							
PGME							
EGME	None		6550				
EEE	None		8407				
EGBE							
Chlorinated							
MDC	None			0.6	8/267		
Chloroform	None		1480a	0.44	8/240		
Carbon tet.	None			1.5	8/140		
1,2-EDC							
1,1,1-TCA							
TCE							
Perk.	48	113	2192	1.9	8/346		
MCB							

Miscellaneous solvents

Solute	Azeotrope X% w/w	°C	Reference	Solute γ^∞	Reference	Partition coefficient	Reference
Ketones							
Acetone	None		5353	1.2	3+4/181		
MEK				1.0	1x/3/1092		
MIBK	60	115	8849	1.0	3b/531		
Cyclohexanone							
NMP							
Acetophenone							
Ethers							
Diethyl ether							
DIPE							
Dibutyl ether							
MTBE							
1,4-Dioxane	None		2769a	1.0	1x/3/1092		
THF							
Esters							
Me acetate							
Et acetate							
i-Propyl acetate							
n-Butyl acetate	None		8850				
Cellosolve acetate							
Miscellaneous							
DMF							
DMAc							
DMSO							
Sulfolane							
CS_2							
Acetic acid	49	138	3139	0.09	5/118		
Aniline	None			0.4	8/500		
Nitrobenzene							
Morpholine							
Pyridine	–		–	–	–	–	–
2-Nitropropane							
Acetonitrile	None		2779b				
Furfuraldehyde							
Phenol	13	183	8842				
Water	57	94	395	2.8	1/469		

2-Nitropropane

Alternative names
2NP, sec nitropropane

Reference codes			
CAS number	79 46 9	Hazchem code	2Y
UN number	2608	EPA code	U171

Physical properties			
Molecular weight	89	Cubic expansion coeff (per °C × 10^3)	1.1
Empirical formula	$C_3H_7N_1O_2$	Surface tension (@20°C dyn/cm)	30
Boiling point (°C)	120	Absolute viscosity (@25°C cP)	0.74
Freezing point (°C)	–93	Refractive index (25°C)	1.392
Specific gravity (20/4)	0.992		

Fire hazards			
Flash point (closed cup °C)	28	Lower explosive limit (ppm)	26000
Autoignition temperature (°C)	428	Upper explosive limit (ppm)	110000
Electrical conductivity	5E–7		

Health hazards			
IDLH (ppm)	2300	Vapour concentration @21°C ppm	22000
OES-TWA	5	Vapour density (relative to air)	3.18
OES-STEL		Vapour pressure @21°C mmHg	16
Odour threshold (ppm)	300	POCP	

Aqueous effluent	
Solubility in water (25°C %w/w)	1.76
Solubility of water in (25°C %w/w)	0.5
Log_{10} activated carbon partition	
Log_{10} partition in octanol/water (w/w)	
Biological oxygen demand w/w (days)	
Theoretical oxygen demand w/w	1.35

Vapour pressure equation constants (Log_{10}, mmHg)		
Antoine equation	A	7.4211
	B	1625.43
	C	237.6
Cox chart	A	
	B	

Solvent properties			
Solubility parameter	10.4	Kauri butanol value	
Dipole (D)	1.9	Evaporation time (ether = 1)	10
Dielectric constant (20°C)	25.5	Evaporation time (BuAc = 1)	1.5
Polarity (water 100)	37.3		

Thermal information	
Latent heat (cal/mol)	8811
Nett heat of combustion (kcal/gmol)	441
Specific heat (cal/mol/°C)	42
Critical pressure (MN/m^2)	
Critical temperature (K)	618
Latent heat of fusion (cal/mol)	
Van der Waals' volume	3.36
Van der Waals' surface area	2.94
Molar volume	90.1

Solute	Azeotrope X% w/w	°C	Reference	Solute γ^{∞}	Reference	Partition coefficient	Reference
Hydrocarbons							
n-Pentane				6.5	1x/1/63		
n-Hexane	3	68	6284	6.6	6a/510		
n-Heptane	21	95	6289	4.0	6b/100		
n-Octane	47	111	6291				
n-Nonane	75	118	6293				
n-Decane							
2,2,4-TMP	21	95	6292				
Cyclohexane	10	81	6283	5.7	1x/1/63		
Benzene	None		6281	1.3	7/186		
Toluene	18	110	6285				
Ethylbenzene	92	120	6290				
Xylenes							
C_9 Aromatics							
Tetralin							
Alcohols							
Methanol	None		1977	8.4	1x/1/63	1.10	V2/86
Ethanol	6	78	3978	8.4	1x/1/63		
n-Propanol	25	96	6271				
i-Propanol	4	82	6270				
n-Butanol	52	112	6275				
i-Butanol	33	105	6278				
s-Butanol	18	99	6276				
n-Amyl alc.	85	120	6280				
i-Amyl alc.							
Cyclohexanol							
1-Octanol							
Ethanediol							
DEG							
1,2-Propanediol							
Glycol ethers							
PGME							
EGME							
EEE	85	119	6279				
EGBE							
Chlorinated							
MDC				0.9	1x/1/63		
Chloroform				0.9	1x/1/63		
Carbon tet.	None		1114	2.3	1x/1/63		
1,2-EDC							
1,1,1-TCA							
TCE							
Perk.							
MCB							

Miscellaneous solvents

Solute	Azeotrope X% w/w	°C	Reference	Solute γ^∞	Reference	Partition coefficient	Reference
Ketones							
Acetone							
MEK	None		6272				
MIBK							
Cyclohexanone	None		6282				
NMP							
Acetophenone							
Ethers							
Diethyl ether							
DIPE							
Dibutyl ether							
MTBE							
1,4-Dioxane							
THF	None		6273				
Esters							
Me acetate							
Et acetate	None		6274				
i-Propyl acetate							
n-Butyl acetate							
Cellosolve acetate							
Miscellaneous							
DMF							
DMAc							
DMSO							
Sulfolane							
CS_2				4.0	1x/1/63		
Acetic acid							
Aniline							
Nitrobenzene							
Morpholine							
Pyridine							
2-Nitropropane	–		–	–	–	–	–
Acetonitrile							
Furfuraldehyde							
Phenol							
Water	71	89	290				

Acetonitrile

Alternative names
Methyl cyanide, ACN, ethane nitrile, cyanomethane

Reference codes

CAS number	75 05 8	Hazchem code	2WE
UN number	1648	EPA code	U003

Physical properties

Molecular weight	41	Cubic expansion coeff (per °C × 10^3)	1.4
Empirical formula	$C_2H_3N_1$	Surface tension (@20°C dyn/cm)	29.1
Boiling point (°C)	81.6	Absolute viscosity (@25°C cP)	0.38
Freezing point (°C)	–44	Refractive index (25°C)	1.342
Specific gravity (20/4)	0.782		

Fire hazards

Flash point (closed cup °C)	6	Lower explosive limit (ppm)	44000
Autoignition temperature (°C)	524	Upper explosive limit (ppm)	160000
Electrical conductivity	6E–10		

Health hazards

IDLH (ppm)	4000	Vapour concentration @21°C ppm	96000
OES-TWA	40	Vapour density (relative to air)	1.42
OES-STEL	60	Vapour pressure @21°C mmHg	71
Odour threshold (ppm)	40	POCP	

Aqueous effluent

Solubility in water (25°C %w/w)	Total
Solubility of water in (25°C %w/w)	Total
Log_{10} activated carbon partition	
Log_{10} partition in octanol/water (w/w)	–0.34
Biological oxygen demand w/w (days)	1.22
Theoretical oxygen demand w/w	2.15

Vapour pressure equation constants (Log_{10}, mmHg)

Antoine equation	A	7.33986
	B	1482.29
	C	250.523
Cox Chart	A	7.12578
	B	1322.7

Solvent properties

Solubility parameter	11.9	Kauri butanol value	
Dipole (D)	3.2	Evaporation time (ether = 1)	2.04
Dielectric constant (20°C)	37.5	Evaporation time (BuAc = 1)	
Polarity (water 100)	46		

Thermal information

Latent heat (cal/mol)	7134
Nett heat of combustion (kcal/gmol)	289
Specific heat (cal/mol/°C)	22
Critical pressure (MN/m^2)	4.83
Critical temperature (K)	548
Latent heat of fusion (cal/mol)	
Van der Waals' volume	1.87
Van der Waals' surface area	1.72
Molar volume	52.86

Solute	Azeotrope X% w/w	Azeotrope °C	Azeotrope Reference	Solute γ^∞	Reference	Partition coefficient	Reference
Hydrocarbons							
n-Pentane	11	35	2792	21.3	1x/3/953		
n-Hexane	28	57	2800	27	1x/1/15		
n-Heptane	46	69	2803	32.8	6b/79		
n-Octane	66	77	2810	57	1x/1/16		
n-Nonane		80	2812	86	1x/1/16		
n-Decane		82	2815				
2,2,4-TMP	41	69	2811	44	1x/1/16		
Cyclohexane	33	62	2797	22.0	1x/1/14		
Benzene	34	73	2795	2.5	7/124		
Toluene	76	81	2801	4.5	7/373		
Ethylbenzene	None		2804	5.0	7/465		
Xylenes	None		2805	5.5	7/499		
C_9 Aromatics				7.3	1x/3/955		
Tetralin							
Alcohols							
Methanol	81	64	1925	3.0	2a/43		
Ethanol	44	73	2760	3.7	2a/298		
n-Propanol	72	81	2768	6.5	1x/3/951		
i-Propanol	52	75	2767	2.4	2f/40		
n-Butanol	None			4.8	2d/156		
i-Butanol	None		2779	3.3	2f/300		
s-Butanol				2.7	2d/241		
n-Amyl alc.							
i-Amyl alc.							
Cyclohexanol							
1-Octanol							
Ethanediol				9.2	2f/1		
DEG							
1,2-Propanediol							
Glycol ethers							
PGME							
EGME				1.7	2d/109		
EEE							
EGBE							
Chlorinated							
MDC	None		1546	1.2	8/258		
Chloroform	None		1433	1.4	8/217		
Carbon tet.	17	65	1095	4.4	8/86		
1,2-EDC	49	79	2757	1.4	8/364		
1,1,1-TCA							
TCE	29	75	2280	3.4	8/349		
Perk.							
MCB	None		2794a	2.8	8/381		

Miscellaneous solvents

Solute	Azeotrope X% w/w	°C	Reference	Solute γ^{∞}	Reference	Partition coefficient	Reference
Ketones							
Acetone	None		2762	1.0	3+4/143		
MEK	27			1.2	3b/268		
MIBK							
Cyclohexanone							
NMP							
Acetophenone							
Ethers							
Diethyl ether	None		2778	3.2	3+4/499		
DIPE							
Dibutyl ether							
MTBE							
1,4-Dioxane	None		2769a	1.4	1x/3/951		
THF							
Esters							
Me acetate	None		2763	1.1	5/354		
Et acetate	23	75	2770	1.6	5/455		
i-Propyl acetate	60	80	2788				
n-Butyl acetate				1.8	5/577		
Cellosolve acetate							
Miscellaneous							
DMF	None		2765				
DMAc							
DMSO							
Sulfolane							
CS_2				17.9	8/320		
Acetic acid	None		2758a				
Aniline				1.0	8/385		
Nitrobenzene							
Morpholine							
Pyridine	None		2779b	1.8	1x/1/14		
2-Nitropropane							
Acetonitrile	–		–	–	–	–	–
Furfuraldehyde							
Phenol							
Water	84	76	226	6.1	1/81		

Furfuraldehyde

Alternative names
Furfural, furfurol, 2-furaldehyde, fural

Reference codes			
CAS number	98 01 1	Hazchem code	
UN number	1199	EPA code	U125
Physical properties			
Molecular weight	96	Cubic expansion coeff (per °C × 10^3)	1.06
Empirical formula	$C_5H_4O_2$	Surface tension (@20°C dyn/cm)	45
Boiling point (°C)	162	Absolute viscosity (@25°C cP)	1.4
Freezing point (°C)	–37	Refractive index (25°C)	1.524
Specific gravity (20/4)	1.160		
Fire hazards			
Flash point (closed cup °C)	62	Lower explosive limit (ppm)	21000
Autoignition temperature (°C)	315	Upper explosive limit (ppm)	193000
Electrical conductivity			
Health hazards			
IDLH (ppm)	250	Vapour concentration @21°C ppm	2400
OES-TWA	2	Vapour density (relative to air)	3.33
OES-STEL	10	Vapour pressure @21°C mmHg	1.81
Odour threshold (ppm)	0.2	POCP	

Aqueous effluent	
Solubility in water (25°C %w/w)	8.4
Solubility of water in (25°C %w/w)	5.0
Log_{10} activated carbon partition	
Log_{10} partition in octanol/water (w/w)	+0.23
Biological oxygen demand w/w (days)	0.77 (5)
Theoretical oxygen demand w/w	1.67

Vapour pressure equation constants (Log_{10}, mmHg)

Antoine equation	A	8.40200
	B	2338.49
	C	261.638
Cox chart	A	
	B	

Solvent properties			
Solubility parameter	11.2	Kauri butanol value	
Dipole (D)	3.6	Evaporation time (ether = 1)	75
Dielectric constant (20°C)	41.9	Evaporation time (BuAc = 1)	
Polarity (water 100)			

Thermal information	
Latent heat (cal/mol)	9216
Nett heat of combustion (kcal/gmol)	539
Specific heat (cal/mol/°C)	36
Critical pressure (MN/m^2)	5.03
Critical temperature (K)	660
Latent heat of fusion (cal/mol)	
Van der Waals' volume	3.17
Van der Waals' surface area	2.48
Molar volume	83.23

Solute	Azeotrope X% w/w	Azeotrope °C	Azeotrope Reference	Solute γ^∞	Reference	Partition coefficient	Reference
Hydrocarbons							
n-Pentane							
n-Hexane							
n-Heptane	5	98	8781	7.1	3+4/50		
n-Octane	None		8797	6.0	3a/137		
n-Nonane							
n-Decane				11.7	3+4/59		
2,2,4-TMP				8.9	3+4/55		
Cyclohexane	None		8763	8.0	3+4/45		
Benzene	None		8760	1.6	3+4/44		
Toluene	None		8776	1.7	3a/135		
Ethylbenzene	None		8783	2.5	3+4/51		
Xylenes	10	139	8785	2.8	3+4/52		
C_9 Aromatics	60	155	8805				
Tetralin							
Alcohols							
Methanol	None			1.0	2c/140		
Ethanol	None			3.8	2a/383		
n-Propanol						0.21	V2/558
i-Propanol						0.78	V2/591
n-Butanol				2.4	2f/155	0.12	V3/115
i-Butanol						0.06	V3/137
s-Butanol							
n-Amyl alc.							
i-Amyl alc.							
Cyclohexanol	5	156	8764				
1-Octanol	None		8789				
Ethanediol	None		4214			0.82	V2/421
DEG							
1,2-Propanediol							
Glycol ethers							
PGME	14	151	8753				
EGME	None		6549				
EEE	None		8406				
EGBE	88	161	8769				
Chlorinated							
MDC				0.9	3a/115		
Chloroform	None		1480	0.9	3+4/36		
Carbon tet.	None		1140	3.6	3+4/35		
1,2-EDC				1.1	3a/119		
1,1,1-TCA							
TCE				2.1	3+4/37		
Perk.	None		2191	2.4	3a/117		
MCB	None		8758				

Miscellaneous solvents

Solute	Azeotrope X% w/w	Azeotrope °C	Azeotrope Reference	Solute γ^∞	Reference	Partition coefficient	Reference
Ketones							
Acetone				2.8	3a/121	0.25	V2/471
MEK						0.11	V3/21
MIBK				1.6	3a/126		
Cyclohexanone	None		8762				
NMP							
Acetophenone							
Ethers							
Diethyl ether							
DIPE							
Dibutyl ether	20	138	8788	6.1	3a/139		
MTBE							
1,4-Dioxane							
THF							
Esters							
Me acetate							
Et acetate	None		7574	1.6	3a/123		
i-Propyl acetate							
n-Butyl acetate				1.6	3+4/46		
Cellosolve acetate	None		8766				
Miscellaneous							
DMF						1.13	V2/539
DMAc							
DMSO							
Sulfolane							
CS_2							
Acetic acid	None		3138			0.34	V2/230
Aniline							
Nitrobenzene							
Morpholine							
Pyridine						0.08	V3/182
2-Nitropropane							
Acetonitrile							
Furfuraldehyde	–		–	–	–	–	–
Phenol	None		8761				
Water	35	98	394	0.8	1/455		

Phenol

Alternative names
Hydroxy benzene, carbolic acid

Reference codes

CAS number	108 95 2	Hazchem code	
UN number	1671	EPA code	

Physical properties

Molecular weight	94	Cubic expansion coeff (per °C × 10^3)	0.9
Empirical formula	$C_6H_6O_1$	Surface tension (@55°C dyn/cm)	36.5
Boiling point (°C)	182	Absolute viscosity (@43°C cP)	4.3
Freezing point (°C)	+41	Refractive index (41°C)	1.542
Specific gravity (41/4)	1.058		

Fire hazards

Flash point (closed cup °C)	79	Lower explosive limit (ppm)	17000
Autoignition temperature (°C)	715	Upper explosive limit (ppm)	86000
Electrical conductivity	2.7E–8		

Health hazards

IDLH (ppm)	100	Vapour concentration @21°C ppm	815
OES-TWA	5	Vapour density (relative to air)	3.26
OES-STEL	10	Vapour pressure @21°C mmHg	0.62
Odour threshold (ppm)	20	POCP	

Aqueous effluent

Solubility in water (25°C %w/w)	8.4
Solubility of water in (25°C %w/w)	28.7
Log_{10} activated carbon partition	4.0
Log_{10} partition in octanol/water (w/w)	+1.47
Biological oxygen demand w/w (days)	
Theoretical oxygen demand w/w	2.38

Vapour pressure equation constants (Log_{10}, mmHg)

Antoine equation	A	6.9305
	B	1382.65
	C	159.5
Cox chart	A	7.84460
	B	2045.1

Solvent properties

Solubility parameter	11.3	Kauri butanol value	
Dipole (D) @40°C	2.2	Evaporation time (ether = 1)	
Dielectric constant (60°C)	10.0	Evaporation time (BuAc = 1)	
Polarity (water 100)	94.8		

Thermal information

Latent heat (cal/mol)	6768
Nett heat of combustion (kcal/gmol)	700
Specific heat (cal/mol/°C)	52
Critical pressure (MN/m^2)	6.13
Critical temperature (K)	694
Latent heat of fusion (cal/mol)	2750
Van der Waals' volume	3.55
Van der Waals' surface area	2.68
Molar volume	83.14

Solute	Azeotrope X% w/w	°C	Reference	Solute γ^∞	Reference	Partition coefficient	Reference
Hydrocarbons							
n-Pentane				10.9	1x/1/231		
n-Hexane				13.6	1x/1/233		
n-Heptane	None		10936	12.8	1x/1/234		
n-Octane	4	125	10959	19.8	2x/382		
n-Nonane				26.8	1x/1/235		
n-Decane	35	168	11016				
2,2,4-TMP				3.0	2b/383		
Cyclohexane				7.1	1x/1/232		
Benzene				2.6	1x/1/231		
Toluene	None		10920	2.8	1x/1/233		
Ethylbenzene	None		10943				
Xylenes	None		10944	3.2	1x/1/234		
C_9 Aromatics				2.0	2b/385		
Tetralin							
Alcohols							
Methanol						0.37	V2/125
Ethanol							
n-Propanol							
i-Propanol						0.13	V2/606
n-Butanol							
i-Butanol							
s-Butanol							
n-Amyl alc.	None		9749				
i-Amyl alc.	None		9822				
Cyclohexanol	87	183	10895	0.3	2b/370		
1-Octanol	13	195	10962				
Ethanediol	22	199	4240	2.6	2d/11		
DEG				0.8	2f/339		
1,2-Propanediol							
Glycol ethers							
PGME	86	183	9962				
EGME	None		6568				
EEE	None		8426				
EGBE	63	186	10904				
Chlorinated							
MDC				1.7	1x/1/230		
Chloroform				1.8	1x/1/230		
Carbon tet.	None		1155	4.2	1x/1/230		
1,2-EDC				2.1	1x/1/230		
1,1,1-TCA							
TCE							
Perk.							
MCB	None		10510				

Miscellaneous solvents

Solute	Azeotrope X% w/w	°C	Reference	Solute γ^{∞}	Reference	Partition coefficient	Reference
Ketones							
Acetone	None		5375				
MEK	None		7370	0.33	2b/358		
MIBK							
Cyclohexanone	72	185	10889	0.11	2b/368		
NMP							
Acetophenone	8	202	10939	0.3	2b/381		
Ethers							
Diethyl ether							
DIPE							
Dibutyl ether	None		10960				
MTBE							
1,4-Dioxane							
THF							
Esters							
Me acetate							
Et acetate							
i-Propyl acetate							
n-Butyl acetate	None		10896	0.46	2b/373		
Cellosolve acetate	72	185	10898				
Miscellaneous							
DMF							
DMAc							
DMSO							
Sulfolane							
CS_2				3.2	1x/1/230		
Acetic acid							
Aniline	42	185	10883				
Nitrobenzene							
Morpholine							
Pyridine	87	183	8842				
2-Nitropropane							
Acetonitrile							
Furfuraldehyde	None		8761				
Phenol	–		–	–	–	–	–
Water	9	99	487	12.5	1/496		

Water

Alternative names

Reference codes

CAS number		Hazchem code	
UN number		EPA code	

Physical properties

Molecular weight	18	Cubic expansion coeff (per °C × 10^3)	0.21
Empirical formula	H_2O_1	Surface tension (@20°C dyn/cm)	72.75
Boiling point (°C)	100	Absolute viscosity (@25°C cP)	0.89
Freezing point (°C)	0	Refractive index (25°C)	1.332
Specific gravity (20/4)	0.998		

Fire hazards

Flash point (closed cup °C)		Lower explosive limit (ppm)	
Autoignition temperature (°C)		Upper explosive limit (ppm)	

Electrical conductivity (×10^{-1} siemen/cm) 5.0 (see *Key to tables*)

Health hazards

IDLH (ppm)		Vapour concentration @21°C ppm	25000
OES-TWA		Vapour density (relative to air)	0.625
OES-STEL		Vapour pressure @21°C mmHg	19
Odour threshold (ppm)		POCP	

Aqueous effluent

Solubility in water (25°C %w/w)
Solubility of water in (25°C %w/w)
Log_{10} activated carbon partition
Log_{10} partition in octanol/water (w/w)
Biological oxygen demand w/w (days)
Theoretical oxygen demand w/w

Vapour pressure equation constants (Log_{10}, mmHg)

Antoine equation	A	8.07131
	B	1730.63
	C	233.426
Cox chart	A	
	B	

Solvent properties

Solubility parameter ($cal^{1/2}$ $cm^{3/2}$)	23.4	Kauri butanol value	
Dipole (D)	1.87	Evaporation time (ether = 1)	
Dielectric constant (20°C)	79.7	Evaporation time (BuAc = 1)	
Polarity (water 100)	100		

Thermal information

Latent heat (cal/mol)	9703
Nett heat of combustion (kcal/gmol)	11.5
Specific heat (cal/mol/°C)	18
Critical pressure (MN/m^2)	22.1
Critical temperature (K)	647
Latent heat of fusion (cal/mol)	1432
Van der Waals' volume	0.92
Van der Waals' surface area	1.40
Molar volume	18.02

Solute	Azeotrope X% w/w	°C	Solubility of solute in X (ppm) @25°C	Solute γ^{∞}	Reference	H atm/mole fraction	Upper CST, °C
Hydrocarbons							
n-Pentane	1.4	35	38	870	1x/4/1656	70250	
n-Hexane	5.6	62	9.5	4500	1x/4/1658	71730	
n-Heptane	13	79	3	11000	1x/4/1659	150000	
n-Octane	25	90	0.6	96100	1x/4/1659	274000	
n-Nonane	40	95	0.2			330000	
n-Decane	51	97	0.02			262000	
2,2,4-TMP	11	79	2.2			186000	
Cyclohexane	8.4	69	55			10700	
Benzene	8.8	69	1800	2150	1x/4/1657	309	306
Toluene	19	85	520	9700	1x/4/1658	353	
Ethylbenzene	33	92	165	24000	1x/4/1659	447	
Xylenes	40	95	200	3630	1x/4/1659	313	
C_9 Aromatics	c.50	96					
Tetralin	80	99					
Alcohols							
Methanol	None		Total	2.18	1/40	0.39	
Ethanol	4.0	78	Total	5.80	1/153	0.45	
n-Propanol	28	88	Total	15.0	1/286	0.51	<–23
i-Propanol	12.6	80	Total	13.7	1/329	0.62	<–23
n-Butanol	42	93	73000	114.1	1/407	0.44	127
i-Butanol	33	90	87000	42.3	1/440	0.35	129
s-Butanol	27	87	198000	24.9	1/420	0.60	110
n-Amyl alc.	54	96	17000	22.7	1x/4/1656	0.68	182
i-Amyl alc.	50	95		60.6	1a/382		
Cyclohexanol	70	98	43000	115.4	1/514		184.7
1-Octanol	90	99	6000			0.88	
Ethanediol	None		Total	0.23	1a/173		<20
DEG	None		Total	2.23	1a/353		<20
1,2-Propanediol	None		Total	0.61	1a/337		<20
Glycol ethers							
PGME	35	97	Total				
EGME	78	100	Total				<20
EEE	87	98	Total	6.9	1a/450		<20
EGBE	79	99	Total*	14.8	1/526		128
Chlorinated							
MDC	1.5	38	13000	336	1/1	138	
Chloroform	2.8	56	8200	665	1x/4/1644	225	
Carbon tet.	4	66	770	6400	1x/4/1644	1634	
1,2-EDC	8.7	72	8100	626	1x/4/1648	65	
1,1,1-TCA	4	65	1300	5500	1x/4/1647	1666	
TCE	6.2	73	1100	5100	1x/4/1646	648	
Perk.	16.5	88	150			1492	
MCB	28	90	490	3000	1x/4/1657	25.2	>220

Miscellaneous solvents

Bibliography

1. *Perry's Chemical Engineers' Handbook*. McGraw-Hill, New York. The most complete collection of information on industrial chemical properties and chemical engineering theory and practice.
2. *Dechema Chemistry Data Series*. Frankfurt.
   a. Volume I. *Vapor–Liquid Equilibrium Data Collection*. Nineteen parts by 1996. A very large collection of binary and ternary vapour–liquid data and activity coefficients for organic materials commonly used in industry. The majority of the VLE of binary mixtures are shown in graphic form. All solvents listed have their Antoine constants included and the values of their Van der Waals surface area and volume.
   b. Volume V. *Liquid–Liquid Equilibrium Data Collection*. Four parts by 1996. The liquid ternary mixtures are shown in graphic form which allow the possibility of extraction of a solute from one solvent to another to be estimated.
   c. Volume IX. *Activity Coefficients at Infinite Dilution*. Four parts by 1996. A very large collection of activity coefficients of liquid solutes in commonly used solvents and stationary phases.
   In all the above 27 books the references cover the original research paper. Each book contains about 550 pages.
3. Riddick, J. A. and Burger, W. B. (1986) *Organic Solvents*, 4th edn. Volume 2 of *Techniques of Chemistry Series*. John Wiley, New York. A good source of physical data and laboratory purification methods for solvents. Hazards and stabilizers/inhibitors for laboratory work are included.
4. Horsley, L. H. (1973) Azeotropic data 111. *Advances in Chemistry Series* No. 116. American Chemical Society, Washington DC. A most comprehensive collection of azeotropes with the original research papers for all azeotropes listed.
5. Weast, R. C. (1975) *CRC Handbook of Chemistry and Physics* (*The Rubber Handbook*). CRC Press, Boca Raton, FL. Although many later editions have been printed those numbered 55 and less contain information on solvent azeotropes which are not included in later editions. However, the later editions contain a large amount of useful information on solvents.
6. Dreisbach, R. R. (1952) *Pressure–Volume–Temperature Relationships of Organic Compounds*, 3rd edn. Handbook Publishers. Contains Cox chart constants and tables of pressure–temperature data for individual compounds.
7. Horvath, A. L. (1982) *Halogenated Hydrocarbons Solubility and Miscibility with Water*. Marcel Dekker, New York. A comprehensive coverage of water-chlorinated hydrocarbons and the mechanism of their miscibility.
8. Flick, F. W. (1993) *Industrial Solvents Handbook*, 4th edn. Noyes Data Corporation, Park Ridge, NJ. A collection of information of the physical properties of commercial solvents, but nothing on their fire and health hazards.
9. De Renzo, D. J. (1986) *Solvents Safety Handbook*. Noyes Data Corporation, Park Ridge, NJ. For every commonly used solvent this book lists the essential physical properties, health data and transport safety information.
10. Sax, N. I. (1984) *Dangerous Properties of Industrial Materials*, 6th edn. Van Nostrand Reinhold, New York. The definitive book on the toxicity of solvents and other organic chemicals.
11. Lide, D. R. (1995) *Handbook of Organic Solvents*. CRC Press, Boca Raton, FL. A compilation of many commonly used solvents with particular stress on analysis and thermodynamic properties.

12. Gallant, R. W. (1993) *Physical Properties of Hydrocarbons*. Gulf Publishing Co. Three volumes. All the properties of the simpler solvents and hydrocarbons up to C_{10} that are required for engineering design purposes plotted over normal working temperature ranges.

13. *Solubility Data Series* (1979 and continuing). Pergamon, Oxford. Over 30 volumes. Very detailed survey of the solubility of materials in water including paraffins, aromatic hydrocarbons, esters and alcohols and of the solubility of water in them.

14. Reichardt, C. (1988) *Solvents and Solvent Effects in Organic Chemistry*, 2nd edn. VCH Verlagsgeseilschaft, Weihheim. Useful for comparing the effects of various solvents and their method of action.

15. Reid, R. C., Prausssnitz, J. M. and Poling, B. E. (1987) *Properties of Gases and Liquids*, 4th edn. McGraw-Hill, New York. Very useful basis for methods of estimating properties that are not available from experiment.

16. Verschueren. *Handbook of Environmental Data on Organic Chemicals*. Van Nostrand Reinhold, New York. A comprehensive collection on the fate of materials in the environment.

17. Hansch, C. and Leo, A. (1979) *Substituent Constants for Correlation Analysis in Chemistry and Biology*. John Wiley, New York. Pomona College $\log_{10} P$ listings.

Solute	Azeotrope X% w/w	Azeotrope °C	Solubility of solute in X (ppm) @25°C	Solute γ^{∞}	Reference	H Atm/Mole fraction	Upper CST, °C
Ketones							
Acetone	None		Total	10.2	1/237	2.38	<–11
MEK	11	73	260000†	27.2	1/363	2.42	139
MIBK	24	88	17000	15.1	1b/337	3.0	
Cyclohexanone	55	96	23000	38.3	1/511	0.82	
NMP	None		Total	1.6	1a/379		
Acetophenone	82	98	5500	1277	1/46		220
Ethers							
Diethyl ether	1.3	34	69000	98.2	1a/257	48.3	202
DIPE	4.5	62	12000	4.3	1/525	96.3	
Dibutyl ether	33	93	300				
MTBE	3	52	43000				
1,4-Dioxane	18	88	Total	8.2	1/382	0.38	<–15
THF	5.3	64	Total‡	24.3	1x/4/1653	6.0	138
Esters							
Me acetate	5	56	245000	23.6	1/264		108
Et acetate	8.5	70	77000	108	1x/4/1653	5.15	
i-Propyl acetate	10	77	29000	242	1x/4/1656		
n-Butyl acetate	29	90	7000	995	1/516		
Cellosolve acetate	50	97	229000				
Miscellaneous							
DMF	None		Total	2.2	1/276		
DMAc	None		Total	1.6	1a/402		<25
DMSO	None		Total	0.43	1/199		
Sulfolane	None		Total	14.8	1a/316		
CS_2	2.8	43		2100	3300	1X/4/1646	1067
Acetic acid	None		Total	3.6	1/102	0.07	<–27
Aniline	41	87	35000	22.6	1/499	0.16	167
Nitrobenzene	86	99	1900			1.03	
Morpholine	None		Total	1.05	1a/327		
Pyridine	43	94	Total	42.8	1/469	0.61	
2-Nitropropane	29	89	17600			6720	
Acetonitrile	15.2	76	Total	32.5	1/81	1.12	–0.9
Furfuraldehyde	65	98	84000	50.6	1/455	2.35	122
Phenol	91	100	84000	44.0	1/496		66

Lower critical solution temperatures

*Butyl cellosolve (DEGBE)	55°C
†Methyl ethyl ketone (MEK)	–6°C
‡Tetrahydrofuran (THF)	72°C